Advances in Green Electronics Technologies in 2023

Edited by Albert Sabban

Published in London, United Kingdom

Advances in Green Electronics Technologies in 2023
http://dx.doi.org/10.5772/intechopen.100759
Edited by Albert Sabban

Contributors
Fernando Vélez Varela, Carlos Daniel Valencia Rincón, Daniel Revelo Alvarado, Hammadi el Farissi, Prasanna Venkatesh Ramesh, Prajnya Ray, Aji Kunnath Devadas, Akshay Surendranp, Tensingh Joshua, Shruthy Vaishali Ramesh, Meena Kumari Ramesh, Ramesh Rajasekaran, Rajashree Lodh, Subhashis Chowdhury, Souvik Chakraborty, Shireen Ibrahim Mohammed, Giovanni Campeol, Lorella Biasio, Silvia Foffano, Davide Scarpa, Giulio Copparoni, Daniel Llarens, Laura Souilla, Santiago A. Masiriz, Gastón R. Lestard, Albert Sabban

First published in London, United Kingdom, 2023 by IntechOpen
IntechOpen is the global imprint of INTECHOPEN LIMITED, registered in England and Wales, registration number: 11086078, 5 Princes Gate Court, London, SW7 2QJ, United Kingdom

British Library Cataloguing-in-Publication Data
A catalogue record for this book is available from the British Library

Additional hard and PDF copies can be obtained from orders@intechopen.com

Advances in Green Electronics Technologies in 2023
Edited by Albert Sabban
p. cm.
Print ISBN 978-1-80356-833-1
Online ISBN 978-1-80356-834-8
eBook (PDF) ISBN 978-1-80356-835-5

For EU product safety concerns:
IN TECH d.o.o., Prolaz Marije Krucifikse Kozulić 3, 51000 Rijeka, Croatia,
info@intechopen.com or visit our website at intechopen.com.

Meet the editor

Dr. Albert Sabban is a communication and radio-frequency specialist for hi-tech companies. He is also a researcher and lecturer in electronics and computers. He received his BS, BSc, and MSc in Electrical Engineering from Tel Aviv University, Israel. He received his Ph.D. in Electrical Engineering from the University of Colorado at Boulder, USA. He was a senior R&D scientist and project leader for more than thirty years at Rafael Advanced Defense Systems Ltd. He developed Radio-frequency Integrated Circuit components on GaAs and silicon substrates as well as microwave components by employing Monolithic Microwave Integrated Circuits, Micro Electromechanical Systems, and Low-Temperature Co-Fired Ceramic technologies. He has also developed wearable communication, 5G, antennas, the Internet of Things (IoT), and medical systems. He is a specialist in microwaves, antennas, computer engineering, electromagnetics, and communication systems. Dr. Sabban led the communication program at Ort Braude College, Israel, from 2013 to 2018. He holds an MBA with a specialization in marketing and management. He developed computer-aided design tools for RF and antenna systems. He has published more than 100 papers and holds several US patents. Dr. Sabban has written and edited ten books on wearable systems, green electronics, and computing technologies.

Contents

Preface

In the last decade, computing devices have become a large part of daily life. Most people have more than one cellular and computing device, thus there are numerous computing networks and devices that consume a huge amount of electrical energy. Designing and manufacturing energy-efficient computing devices, servers, and laptops can help minimize air and water pollution. This is called green computing, a concept born in 1992 with the launch of the Energy Star program in the United States. One of the main goals of electrical and computer engineers should be to evaluate and manufacture green computing devices and servers. Moreover, disposing and recycling of unwanted computing instruments and electronic devices is an important task in creating a green planet. Every person can contribute to green computing via the environmentally responsible use of computing devices and cellular phones.

Green computing may be considered as the study, development, design, engineering, research, production, use, and disposal of computing devices and networks to reduce environmental hazards, energy consumption, and environmental pollution. Computer researchers, designers, developers, manufacturing companies, and vendors are investing in developing green computing networks and modules by reducing the use of hazardous materials and using improved recycling processes for computing modules and devices. Green computing is known as green information technology (green IT).

In the last decade, the world has suffered from severe changes in temperature and climate as well as pollution of ecosystems. Many countries are experiencing extreme heat, severe droughts, depletion of groundwater reserves, and environmental pollution. Climate change is also contributing to the rapid spread of diseases and viruses and the extinction of animal species. Unfortunately, these changes are almost irreversible. One way to mitigate these drastic environmental changes is via green computing. Dangerous chemicals and toxins that pollute the environment are part of computers and electronic waste. These dangerous chemicals can be found in crops and cannot be removed from food products. Polluted air and water and climate changes affect the health of all individuals. Birds, fish, and other animals consume plastic waste, and these plastic particles are making their way into the food chain. Green computing is one way to mitigate these negative effects. Electronic waste management, recycling, green electronic devices, and renewable energy are important challenges in green technologies.

The book presents new topics and innovations in green computing technologies that can be applied in green computing and electronic industries. It is organized into four sections: "Innovation in Renewable Energy for Green Computing", "Green Energy for Computing Networks", "Green Healthcare and Computing Systems", and "Waste Management for Green Computing Systems".

Section 1 includes the intro chapter, “Introductory Chapter: Advances in Green Electronics Technologies in 2023”, Chapter 2, “Green Wearable Sensors for Medical, Energy Harvesting, Communication, and IoT Systems”, Chapter 3, “Variable Renewable Energy: How the Energy Markets Rules Could Improve Electrical System Reliability” and Chapter 4, “Management Methods of Energy Consumption Parameters Using IoT and Big Data”.

Section 2 includes Chapter 5, “From Photovoltaic to Agri-Natural-Voltaic (ANaV)” and Chapter 6, “Valorization of Forest Waste for the Production of Bio-oils for Biofuel and Biodiesel”.

Section 3 includes Chapter 7, “Going Green in Ophthalmic Practice”.

Finally, Section 4 includes Chapter 8, “Analysis of Rainwater Harvesting Method for Supply of Potable Water: A Case Study of Gosaba, South 24 Pargana, India” and Chapter 9, “E-Waste Management in Different Countries: Strategies, Impacts, and Determinants”.

Albert Sabban
Department of Electrical Engineering,
Ort Braude College,
Karmiel, Israel

Section 1

Innovation in Renewable Energy for Green Computing

Chapter 1

Introductory Chapter: Advances in Green Electronics Technologies in 2023

Albert Sabban

1. Introduction

Green computing was started in 1992 and was named Energy Star. In the last decade, computing devices have become a large part of our daily life. Every person has more than one cellular and computing device. There are on our planet milliards of computing networks and devices that consume a huge amount of electrical energy. By designing and manufacturing energy-efficient computing devices, servers, and laptops, we can minimize air and water pollution. One of the main goals of electrical and computer engineers should be to evaluate and manufacture green computing devices, servers, laptops, and tablets. Moreover, disposing and recycling unwanted computing instruments and electronic devices is important in creating a green planet. Every person can contribute to green computing by environmentally responsible use of computing devices and cellular phones.

1.1 Personal contribution to green computing

- Everyone can buy energy-efficient laptops, computers, and other computing devices. Use computing devices that have been awarded a valid certificate of high energy efficiency.
- Everyone can turn off his laptop and computer when not using the computing device.
- Turn the brightness down of the computing devices when they are not being used. Wireless devices, computing devices, tablets, and laptops consume electrical energy in standby mode. Leaving them on but not in use will still drain the device's battery.
- Printers consume a lot of electrical power. Print when necessary and recycle papers. Everyone can employ digital correspondence through file-sharing services and email.
- By using updated software and drivers, we can reduce the electrical power consumption of computing devices. This will ensure that the computing devices

run at optimal energy efficiency and reduces the time and power required to complete computing tasks.

- Recycling of computing devices and electronic products contributes to the green environment.

Green computing may be considered the study, development, design, engineering, research, production, use, and disposal of computing devices and networks to reduce environmental hazards, energy consumption, and environmental pollution. Computer research, designers, developers, manufacturing companies, and vendors are investing in developing green computing networks and modules by reducing the use of hazardous materials and using improved recycling processes for computing modules and devices. Green computing is known as green information technology (green IT).

In 1992 EPA, the Environmental Protection Agency, started the Energy Star project that initiated green computing research and practice.

Green computing main topics and initiatives:

- **Energy Consumption**: Minimizing the electrical energy consumption of computing networks and other computing devices and employing them in an eco-friendly manner.
- **Green disposal:** Disposing and recycling, unwanted computing modules, laptops, tablets, and other computing devices.
- **Green development and design:** Evaluating and designing energy-efficient computing networks and devices, servers, printers, and other computing devices. Energy harvesting technologies minimize the electrical energy consumption of computing networks and other computing devices.
- **Green manufacturing:** Recycling computing devices, computing modules and electronic components, and cellular phones during the manufacturing of computing networks, laptops, and their peripheral devices. Minimizing waste during manufacturing computing networks, computers, and other computing subsystems to minimize environmental pollution because of these activities.

The book presents new topics and innovations in green computing technologies that can be applied in green computing and electronic industries. The book presents the major topics in green computing and electronic technologies such as renewable energy, green energy, green healthcare, waste, and green computing.

In 2023 Green Computing should be used to protect the world from air and water pollution. In the last decade, the world suffered from severe changes in temperature and climate, air, and water pollution, ocean pollution, and sea pollution. Many countries suffer from severe droughts, seawater pollution, seawater levels rise because of the earth's temperature rise, and severe droughts cause depletion of groundwater reserves, sea pollution, air pollution, and river pollution. Climate changes cause the rapid spread of diseases, viruses, and the extinction of animal species. These changes are almost irreversible. The computing and electronic industry, trucks and cars industry, and the cellular phones industry in the last decade depleted and ruined the universe's natural resources. Dangerous chemicals and toxins that pollute air, water, seawater, and groundwater are part of computers and electronic waste.

These dangerous chemicals cannot be removed from food products such as fruit, fish, and vegetables. These dangerous chemicals can be found in food crops, crops grown on polluted soil, fruits, vegetables, fish, and meat. Polluted air, water, and climate changes affect grownups and children's health. Dangerous chemicals and plastic waste kill our world's ocean and sea habitats. Chemical toxins, plastic waste, and other hazardous materials kill birds, fish, and other animals. Fishes and other animals swallow plastic waste, and these creatures become contaminated, and plastic particles are making their way into our daily food chain. The contaminated animals, fishes, and other creatures are served in our food and reach our bodies. Green computing networks, electronic waste management, recycling, green electronic devices, and renewable energy are important challenges and important topics in green technologies.

2. Green computing networks and green electronics technologies

Green computing has an important positive impact on protecting the environment. The information and communication technology industries are responsible for 2% to up to 4% of global greenhouse gas emissions. Due to this fact, green electronics and computing engineering have been in continuous growth in the last decade. Moreover, cellular communication, communication, electronics, and computing industries are facing increasing pressure from governments and legislative institutes to remove toxic and hazardous materials from their products. There is a worldwide environmental movement to use green energy, green materials, and green components in the production of communication, electronic devices, and computers. The energy demands and carbon emissions of computing and the entire communication systems must be dramatically reduced to slow climate changes and to avoid catastrophic environmental disaster. The use of toxic and hazardous materials such as cooper, lead, plastic materials, and other toxic materials is limited or not permitted to minimize pollution and to improve recycling process. Green computing seeks to reduce the carbon emission price tag. When computing system designers take steps to reduce the amount of energy that each module uses to reduce the amount of heat those modules produce, the carbon price tag of computing systems may be reduced considerably. Moreover, computing system designers should increase the life cycle of computing devices and components, so they do not need to be replaced frequently. Computing system designers should increase users' ability to reuse computing products and design recyclable components and modules when they should be replaced.

2.1 Main objectives and activities in green computing and electronics

- Employing Green renewable energy.
- The key goal of green computing is to reduce energy consumption and carbon emission.
- Using efficient computing networks and minimization of energy consumption.
- Using green materials and avoiding hazardous materials.
- Developing proper algorithms to improve the computer's efficiency

- Taking care of computer and electronic waste
- Recycling computers and electronic waste

2.1.1 How to achieve green computing

By applying the following steps green computing and energy efficiency in data centers and other IT centers may be improved.

- Install building green environment systems that are energy efficient, such as solar and renewable energy.
- Install fans throughout computing systems racks to reduce heat.
- Install overhead lighting with low energy consumption with timers or motion detectors to control light switches and reduce the time lights that is used.
- Buy and use energy-efficient servers, laptops, switches, desktop systems, printers, and scanners and other electronic and communication equipment.
- Install energy-efficient windows and doors that have reflective glass to reduce heat.
- Turn off computing systems that are not performing scheduled work.
- Use refillable computing systems and devices.

3. Green computing services and cloud storage

3.1 Green cloud computing services

Cloud computing is an on-demand internet computing service that provides shared computer facilities, data storage to computing devices, servers, and computer processing facilities. Cloud computing is based on sharing computing facilities such as servers, computer networks, data storage devices, computing applications, and other services. This service provides access to a shared pool of configurable computing systems such as servers, computers, data storage computing devices, and other services. Cloud computing services can be rapidly provisioned and released with minimal management effort. Cloud computing and storage solutions provide users and organizations with several capabilities to store and process their data in private cloud computing data centers. Cloud computing allows organizations and companies to get faster computing services and minimize high infrastructure costs such as expensive software costs and servers. Cloud computing allows companies to get their applications up and run with better manageability and less computing maintenance costs. Cloud computing information teams can easily adjust computing resources to perform unpredictable computing business tasks. Cloud computing applies high-performance computing power to perform tens of trillions of computations per second. By using cloud computing services, companies and organizations can focus on their core activities instead of spending money, time, and staff on computer facilities.

Cloud computing and storage solutions cut companies' and organizations' energy consumption.

3.2 Cloud computing storage

Cloud storage provides companies with greener computing services. Cloud computing storage cuts companies' and organizations' energy consumption and computing expenses. By using cloud storage, purchasing additional storage capacity is minimized and storage maintenance tasks and expenses are reduced. Cloud storage is a service package in which data is stored, managed, backed up remotely, and made available to users over a network and internet services. Cloud storage is based on a virtualized infrastructure with accessible interfaces. Cloud-based data is stored in servers located in data centers managed by a cloud provider. A file and its associated metadata are stored in the server by using an object storage protocol. The server assigns an identification number, ID, to each stored file. When the file needs to be retrieved, the user presents the ID to the system, and the content is assembled with all its metadata, authentication, and security. The most common use of cloud services is cloud backup, disaster recovery, and archiving of infrequently accessed data. Cloud storage providers are responsible for keeping the data available and accessible and the physical environment protected and running. People and organizations buy or lease storage capacity from the providers to store and archive data files. Cloud storage services may be accessed via cloud computers and web services that use application programming interfaces, API, such as cloud desktop storage and cloud storage gateways. There are three main cloud-based storage architecture models public, private, and hybrid.

3.3 Importance of cloud computing storage

- Cloud storage provides companies with greener computing services.

- Cloud computing storage cuts companies' and organizations' energy consumption and computing expenses. Cloud storage provides users with immediate access to a broad range of resources and applications hosted in the infrastructure of another web service interface.

- Cloud computing storage can provide the benefits of rapid deployment, greater accessibility, strong protection for data backup, archival and disaster recovery purposes, and reliability.

- Cloud computing storage uses at least two backup servers located in different places around the globe. Cloud computing storage is used as a natural disaster-proof backup.

- Cloud computing storage may be used for copying virtual machine images from the cloud to a desired location or to import a virtual machine image from any designated location. Cloud storage services provide data protection and storage availability. Additional technology computing efforts and cost to ensure data availability and protection of data storage are eliminated.

3.4 Drawbacks of cloud computing storage

- By using cloud computing storage, the security level of the stored data is decreased.
- By using cloud computing storage, the risk of unauthorized physical access to the data is increased.
- By using cloud-based architecture, data is replicated and moved frequently, so the risk of unauthorized data recovery increases drastically.
- Cloud computing storage increases the number of networks that the data is transferred over them.
- Data stored on a cloud requires a wide area network.
- Data privacy may decrease because cloud storage companies have many customers and thousands of servers. Therefore, computing staff may access almost all the data at the entire facility. Encryption keys that the service user keeps limiting the access to data by service provider employees. An amount, of keys, have to be distributed to users via secure channels for decryption. The keys should be securely stored and managed by the users in their devices. Storing these keys requires expensive secure storage.
- Cloud computing storage is a rich resource for both hackers and national security agencies. The cloud store data from many different users and organizations. Hackers see it as a valuable target.
- Cloud storage companies faced lawsuits from the owners of the intellectual property uploaded and shared on the site. Piracy and copyright problems may be enabled by sites that permit file sharing.
- Cloud storage companies can go bankrupt or be purchased by larger foreign companies, and they may suffer from an irrecoverable disaster.

4. Innovation in green renewable energy 2023

Examples of green renewable energy are solar energy, water energy, wind energy, and biological fuel. Natural resources like light, waterfalls, wind, and electromagnetic energy, are used to produce green renewable energy. According to the International Energy Agency, IEA, main markets like the United States, China, and Europe will help solar power production to reach 125GW capacity per year from now until 2025.

4.1 Solar energy

The sun is an infinite source of light and solar energy. Solar panels and cells convert solar energy from natural light into electrical energy through photovoltaics. Using solar energy to generate electricity minimizes the consumption of coal, fuel, and gasoline. Using solar energy results in a huge reduction in air, water, and

Figure 1.
Thin solar panels on a roof.

environmental pollution. **Figure 1** presents thin solar panels on a roof. Innovative low-cost ways solar energy is being used to enhance our daily lives are presented below. Streetlights – The sun charges the batteries during the day, which then powers light-emitting diodes (LEDs) at night to illuminate the streets. Several cities are incorporating smart sensors into streetlights that can even direct drivers to open parking spaces and help first responders during emergency situations. More cities across the world are powering streetlights with solar energy. Using internet-linked sensors with solar-powered streetlights saves both time and money.

4.1.1 Cell phone charger

Portable solar panels, the size of a tablet, can charge cell phones, GPS trackers, tablets, and laptops. USB cell phone chargers can charge a phone to almost full after 2 hours of exposure to solar energy. They can be hooked on backpacks to collect UV sunlight as we walk, making them ideal for outdoor activities.

4.1.2 Vaccine refrigerators

Private companies have been manufacturing solar-powered vaccine refrigerators so healthcare workers in remote areas such as Africa can administer critical medication to those who need it. In developing countries, 24-hour electricity is not guaranteed, and in many cases, there is no electrical grid. This technology solution saved lives for more than 30 years.

4.1.3 Ovens

Solar ovens, called solar cookers, reflect the sun's energy to cook food. Solar ovens can be parabolic or square lined with a reflective material that directs the solar energy into the box, which heats the food evenly. The lid on top is typically made from glass to better focus the sun's energy. They are ideal for off-grid living and are used in developing countries and in outdoor activities. These ovens reduce air pollution that results from burning fuel.

Advantages of Green Solar Energy

- Solar energy is environmentally friendly and does not emit greenhouse gases.
- Solar energy is a clean green energy. Solar energy does not result in the destruction of forests and the environment that occurs with many fossil fuel operations.
- Solar energy offers decentralization of power.
- Solar energy does not rely on constantly mining raw materials such as coal and fuel.
- Solar cells and panels produced today carry up to 30 years of warranty.
- Solar light is a free natural resource and is not degradable.
- Solar energy does not require expensive raw materials like coal, oil, or gasoline. Production of solar energy requires drastically lower operational labor than conventional power production.
- Oil, gasoline, and coal used to produce conventional electricity should be constantly extracted, refined, and transported to the power plant.
- Oil, gasoline, and coal, used to produce electricity, are often transported cross-country or internationally. This transportation has additional costs, including monetary costs, and pollution costs of transport. These costs are avoided with solar energy.
- If the organization, house, or company has no connection to the government electricity facility, it is called an off-grid connection. In this case, the house, organization, facility, company, or community relies only on solar energy. The ability to produce electricity off the grid is a major advantage of solar energy for isolated communities, facilities, and rural areas. "On-grid" means if the house remains connected to the government electricity facility, it is called on grid connection. Solar energy can be produced on or off the grid. Installing power long power lines is significantly difficult and very expensive in these rural areas.
- The sun is a free unlimited commodity that can be sourced from several locations. One of the major advantages of solar energy is the ability to avoid the linkage between political status and energy prices. However, in the fuel markets, there is a strong linkage between political status and energy prices. Solar energy prices are less affected by price manipulations of political status, war, and international relations. Politics status manipulations doubled the price of fuel in the past two decades.

Disadvantage of Solar Energy

- Solar energy cannot be produced at night or in foggy areas. Depending on the weather and sunlight.

- Solar energy installation and maintenance are expensive.
- Solar energy panels consume large area zones.
- Solar energy development and production are expensive.
- The efficiency of solar energy production is relatively low compared to traditional electricity production. The efficiency of solar cells and panels is less than 40%.
- Solar electricity storage technology is expensive and is more appropriate for small houses.

4.2 Wind energy

Wind energy is a form of renewable energy that has gained significant attention in the last 30 years as a potential solution to the world's increasing energy demands. The natural power of the wind can be used to generate electricity without producing harmful emissions like fuel and other toxic energy sources. Cumulative wind capacity worldwide increased from around 20,000 megawatts to more than 500,000 megawatts in the last 15 years. International green organization efforts to minimize climate changes, such as the 2015 Paris treaty, green renewable energy is continuously growing. Wind energy is a popular renewable source of energy that has a minor impact on the environment compared to producing energy by using coal or fuel. Wind power uses the strong wind flow to provide mechanical power through wind turbines to turn electric generators to produce electrical power. Wind kinetic power is used to operate electric turbines and windmills. Windmills cannot be operated in residential areas. Wind energy farms are usually located off-shore or on high mountains. When the wind blows, the turbine's blades spin clock-wise, capturing energy. The main shaft of the wind turbine, connected to a gearbox within the nacelle, is triggered to spin when the wind blows. The gearbox sends the wind energy by the gearbox to the generator. Wind power is converted to electricity. Offshore wind turbines provide steady, reliable clean energy in many countries.

(a) (b)

Figure 2.
(a) Country wind energy farm (b) offshore wind energy site.

Figure 2 presents wind energy turbines. **Figure 2a** presents a country wind energy farm. **Figure 2b** presents an offshore wind energy site. The main advantage of wind energy, in several countries, is that wind is a clean, reliable free source of renewable energy with no air or water pollution. Wind energy turbines are ugly and noisy. A disadvantage of wind energy is that the wind turbines rotating blades kill birds, bats, and other protected birds.

Advantage of Wind Energy

- Wind energy is a green and clean energy that does not directly emit CO_2 or greenhouse gases.

- Wind energy is plentiful, readily available, and does not deplete our valuable natural resources.

- Not degradable.

- Wind power is cost-effective in many regions. Wind energy is cheap.

- Wind energy offers decentralization of power.

- Wind energy is environmentally friendly. Wind energy does not rely on constantly mining raw materials. Wind energy does not result in the destruction of forests and ecosystems that occurs with many fossil fuel operations.

- Wind is a free natural resource.

- Wind energy does not require expensive and ongoing raw materials like oil, coal, or gas. Wind energy requires significantly lower operational labor than conventional power production. Raw materials should not be constantly extracted, refined, and transported to the power plant.

- The wind is an unlimited free commodity that can be sourced from several locations. One of the major advantages of wind energy is the ability to bypass the linkage between political status and energy price. However, in the fuel markets, there is a strong linkage between political status and energy prices. Wind energy prices are less affected by price manipulations of political status, war, and international relations. However, political status manipulations doubled the price of fuel in the past 50 years.

- Oil, coal, and gas used to produce conventional electricity are often transported cross-country or internationally. This transportation has additional costs, including monetary costs, and pollution costs of transport. These costs are avoided with solar energy.

Disadvantage of Wind Energy

- Depend on the weather and wind velocity.

- Cannot be operated in a residential area.

- Consume a large area and is noisy. Moreover, dangerous to birds, bats, and animals.
- Development and production of wind energy sites are expensive, and land is expensive.

4.2.1 Innovation in wind energy – Vertical axis wind turbines

Vertical axis wind turbines (VAWTs) are a new type of wind turbine. Unlike the traditional horizontal-axis wind turbines used for several decades, vertical-axis wind turbines have a unique design that allows them to capture wind energy from any direction. The blades of VAWTs are positioned vertically and rotate around a central axis, making them ideal for use in urban areas and other locations where wind direction is unpredictable. VAWTs have several advantages. Because of their compact design, VAWTs can be installed in various settings, including residential homes and commercial buildings, making them a viable option for decentralized energy generation. Additionally, VAWTs are generally quieter than their horizontal axis turbines, making them better for noise-sensitive environments. Researchers are investigating new materials and designs that could make VAWTs even more efficient and cost-effective. **Table 1** presents the global wind farms database in May 2023.

4.3 Green hydropower, water energy

In water energy sites, the water flow kinetic energy is converted to electric energy. Waterfalls and fast-running water flow may be used to produce electric energy. In the late 19th century, hydropower became a source for generating electricity. The first

Region	Subregion	Operating	Construction	Preconstruction
	Global Total	**786,304**	**163,957**	**967,006**
Africa	Northern Africa	3302	3498	21,880
	Sub-Saharan Africa	4709	949	8573
Americas	Latin America and the Caribbean	41,483	13,004	95,869
	Northern America	152,491	16,896	55,628
Asia	Central Asia	648	798	1240
	Eastern Asia	317,150	80,415	346,178
	South-eastern Asia	7304	2390	54,507
	Southern Asia	30,798	4011	14,115
	Western Asia	12,618	1155	4694
Europe	Eastern Europe	16,522	1219	13,616
	Northern Europe	58,721	17,095	195,754
	Southern Europe	48,519	6227	71,538
	Western Europe	79,814	6562	28,930
Oceania	Australia and New Zealand	12,207	9676	54,483

Table 1.
2023 global wind offshore farms database in mega Watts.

commercial hydroelectric power plant was built at Niagara Falls in 1879. In 1881, streetlamps in the city of Niagara Falls were powered by hydropower. Hydroelectricity can be used to store energy in the form of potential energy between two areas with different heights with pumped-storage hydroelectricity. Water is pumped uphill into cities during periods of low demand. This energy can be released to generate energy when demand is high.

In this decade, hydropower has an important role in the transition to green energy through its massive quantities of low-carbon electricity and its unmatched capabilities for providing flexibility and storage. Many hydropower plants can ramp their electricity generation up and down very rapidly compared with other power plants such as nuclear, natural gas, and coal. This makes sustainable hydropower an attractive foundation for integrating greater amounts of wind and solar power, whose output can vary depending on factors like the weather and the time of day or year. Global hydropower capacity is expected to increase by 17% between 2021 and 2030. In 2020, hydropower supplied around 17% of global electricity generation, making it the single largest energy source of low-carbon power. Hydropower output has increased by 70% over the past 20 years, but its share of the global electricity supply has held steady because of the increases in wind, solar PV, natural gas, and coal. Nonetheless, hydropower currently meets most of the electricity demand across 28 different emerging and developing countries, with a total population of 800 million.

4.4 Energy harvesting

As presented in **Figure 3**, the energy harvesting unit consists of an antenna, rectifying diode and circuit, and a rechargeable battery. **Figure 3a** presents a Circular polarized wearable active receiving sensor with energy harvesting unit. The energy harvesting unit and the radiating element provide a self-powered device. The rectifier diode converts electromagnetic energy, AC energy, to direct current (DC energy). Two types of diode rectifiers are usually used a half wave rectifier or a full wave rectifier, [1–6]. A half-wave rectifier converts only the half cycle of the positive voltage. It allows to harvest only one half of the electromagnetic wave. The efficiency of the half wave rectifier is around 41%. The bridge full wave diode rectifier circuit converts electromagnetic energy to DC energy. The bridge rectifier consists of four diodes D1 through D4. The rectifier output DC voltage, $V_{ODC} = 2V_m/\pi$, The full wave rectifier efficiency is around 81%. Electromagnetic power amount in public centers, stadiums, hospitals, and malls may range from 1 to 5 mW/cm^2. The harvesting system efficiency increases as function of the RF power collected by the harvesting system as listed in **Table 2**. The amplifier amplifies the input power collected by the energy harvesting system and improves the efficiency of the harvesting system. Results listed in **Table 2** are also presented by companies that manufacture RF energy harvesting systems, see [1–7]. If the RF radiating sources are close to the harvesting system, the RF power collected by the harvester will be higher. A wearable medical system with energy harvesting unit is presented in **Figure 3b**.

The continuous growth in the production of portable RF communication systems and cellular phones increases the consumption of electrical energy and batteries. In the last 30 years, the trend to use green free space energy, such as light, electromagnetic energy, heat, vibration, muscle motion, and other energy-green sources, has become very important, attractive, and useful. Several methods, research, and inventions to produce electricity from free green energy sources have been published, see [1–7]. Electromagnetic harvesting may be useful to recharge phones and batteries only if we collect electromagnetic energy that can charge electronic devices. Energy harvesting units can eliminate the need to charge batteries by using electrical cables. It is crucial to harvest electromagnetic energy from many transmitting RF modules and systems. Multiband wideband antennas should be employed to harvest as much RF energy as possible. Due to low RF energy densities in free space, wideband efficient

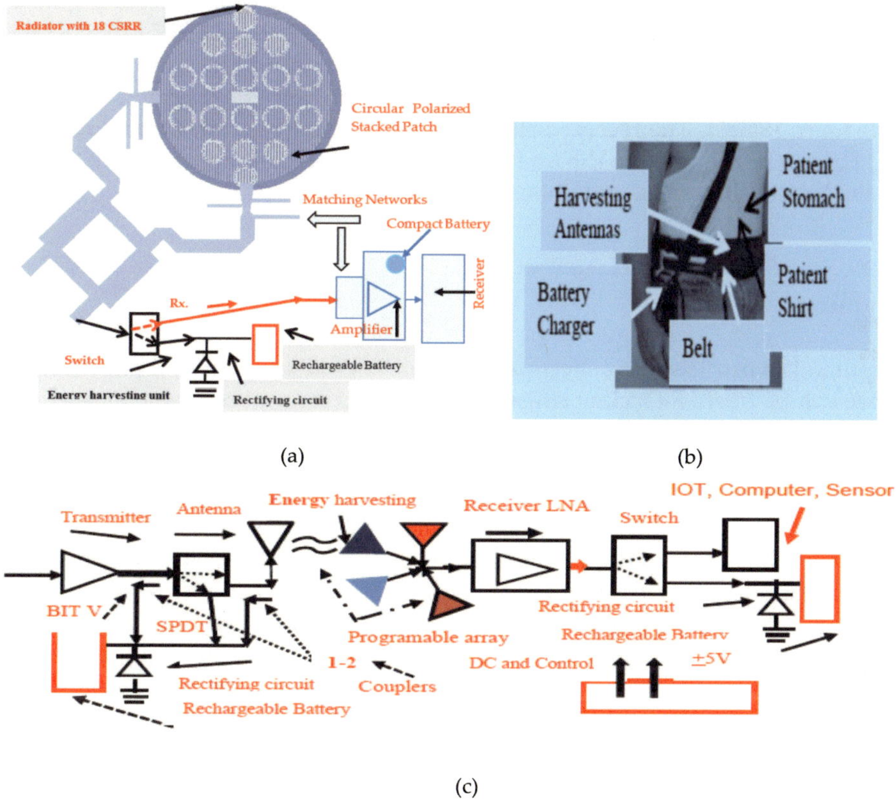

Figure 3.
Sensors with RF energy harvesting unit. (a) Circular polarized active receiving sensor with energy harvesting unit. (b) Wearable medical system. (c) Dual mode energy harvesting concept.

Input power dBm	Efficiency %	Remarks
−4 to −6	8–12	Low efficiency
−3 to −2	28–32	Low efficiency
−1 to +1	48–52	Good efficiency
2–3	52–56	Good efficiency
4–5	52–56	Good efficiency
6–8	56–58	Good efficiency
9–11	60–65	Best efficiency

Table 2.
Measured harvester efficiency as function of input collected power.

antennas should be developed. The antennas should radiate efficiently at a specific frequency range and polarization. To meet the RF system requirements, the antenna radiation pattern should have a wide beam width. Printed antennas were used to harvest RF energy as presented in the literature, [4–10]. RF energy harvesting systems collect electromagnetic waves propagating in the air. This RF energy is stored and used to recharge phones, batteries, and other electrical modules. In the last 20 years, there has been a huge increase in the amount of electromagnetic energy in the air. The electromagnetic energy harvesting system operates as a Dual Mode harvesting unit. The Low Noise Amplifier, LNA, is part of the RF system. The LNA DC bias voltages are supplied by the DC unit of the RF system. The programmable array, shown in **Figure 3c**, consists of antennas that can harvest electromagnetic energy from around 0.15GHz up to 18GHz. The received electromagnetic energy is transformed into DC energy. The energy coupled to the built-in test port, −20 dB, may be used to recharge electronic devices, batteries, and phones.

5. Recycling

Recycling is a process that collects and processes used products and materials that otherwise can be trash. Recycling conserves raw materials and components and saves the additional energy that companies would use to produce new devices from scratch. The rapid growth in manufacturing cellular wireless communication systems, tablets, and laptops over the last 20 years has resulted in most of the universe's population owning computers, smartphones, laptops, smart watches, I-pads, and other computing devices. In 2023, the number of unwanted computing devices and electronic devices is huge. With this big number of computing devices and phones being produced and discarded, a new environmental disaster strikes our world. Electronic and computing waste is filling up landfills and trash storage areas. This computing and electronics waste contains toxic, hazardous materials that pollute the environment and danger the health of human beings and animals. This electronics and computing waste increase air, water, and environmental pollution. Recycling computing devices, cellular phones, and laptop waste, used batteries, plastic waste, and other hazardous devices drastically decreases environmental pollution. Recycling saves original raw materials and reduces waste and pollution.

6. Challenges in green computing technologies in the next three years

We should encourage companies to use green renewable energy, green materials, and green components in the development and production of smartphones, electronic devices, and other computing devices.

We should encourage companies, organizations, and global activities to reuse and recycle computing devices, electronic waste, and old smartphones. In green computing and green electronic industries, the use of hazardous and toxic materials such as copper, lead, plastic materials, and other toxic materials should be limited to decrease air, water, and other environmental pollution. Legislators should encourage and force companies to produce green computing devices. High-Tech companies and computing device manufacturers should be encouraged to develop and produce electronic devices that are green and environmentally friendly.

6.1 Green computing and electronics technologies challenges and innovations

- Developing efficient renewable and harvesting energy for the electronic and computing industry.
- Improving the recycling process and using recycled components and devices in the manufacturing of computing and electronic devices.
- Design and development of new green materials, green computing, and electronic devices.
- Developing efficient and cheap green solar energy, wind energy, and hydropower energy.
- Developing methods to improve the Recycling process. Recycling most of the computing and electronics waste product. Developing new green plastic and packages.
- Development and production of green electrical cars, motorcycles, and airplanes.
- We should encourage legislators, leaders, people, young people, and kids to lead around the globe activities to decrease air pollution, water pollution, and climate change.

Author details

Albert Sabban
Department of Electrical Engineering, Ort Braude College, Karmiel, Israel

*Address all correspondence to: avi_111@netvision.net.il

References

[1] Paradiso JA, Starner T. Energy scavenging for mobile and wireless electronics. IEEE Pervasive Computing. 2005;**4**(1):18-27

[2] Valenta C, Durgin GD. Harvesting wireless power: Survey of energy-harvester conversion efficiency in far-field, wireless power transfer systems. IEEE Microwave Magazine. 2014;**15**(4): 108-120

[3] Nintanavongsa P, Muncuk U, Lewis DR, Chowdhury KR. Design optimization and implementation for RF energy harvesting circuits. IEEE Journal on Emerging and Selected Topics in Circuits and Systems. 2012; **2**(1):24-33

[4] Devi KA, Sadasivam S, Din NM, Chakrabarthy CK. Design of a 377 Ω patch antenna for ambient RF energy harvesting at downlink frequency of GSM 900. In: Proceedings of the 17th Asia Pacific Conference on Communications (APCC'11). Sabah, Malaysia; 2011. pp. 492-495

[5] Rahim R, Malek F, Anwar SFW, Hassan SLS, Junita MN, Hassan HF. A harmonic suppression circularly polarized patch antenna for an RF ambient energy harvesting system. In: Proceedings of the IEEE Conference on Clean Energy and Technology (CEAT'13). Lankgkawi, Malaysia: IEEE; 2013. pp. 33-37

[6] Krakauskas M, Sabaawi AMA, Tsimenidis CC. Suspended patch microstrip antenna with cut rectangular slots for RF energy harvesting. In: Proceedings of the 10th Loughborough Antennas and Propagation Conference (LAPC'14). Loughborough, UK; 2014. pp. 304-307

[7] Sabban A. Low Visibility Antennas for Communication Systems. USA: Taylor & Francis Group; 2015

[8] Sabban A. Wideband RF Technologies and Antenna in Microwave Frequencies. USA: Wiley Sons; 2016

[9] Sabban A. Wearable Communication Systems and Antennas for Commercial, Sport, and Medical Applications. England, GB: IOP Publication; 2018

[10] Sabban A. Novel Wearable Antennas for Communication and Medical Systems. Boca Raton, FL, USA: Taylor & Francis Group; 2017

Chapter 2

Green Wearable Sensors for Medical, Energy Harvesting, Communication, and IoT Systems

Albert Sabban

Abstract

This chapter presents novel passive and active wearable sensors for biomedical systems, energy harvesting, and communication devices. Design tradeoffs, simulation, and measured results of compact efficient sensors for communication, energy harvesting, IoT, and healthcare systems are discussed in this chapter. The new sensors are green sensors with an energy harvesting unit. The sensor electrical parameters near the human body were evaluated by employing RF CAD software. The sensors are flexible passive and active devices with high efficiency and low cost. Low-cost sensor may be developed by printing the printed antenna with the antenna feed network and the active components on the same board. Efficient metamaterial sensors were developed to improve the system electrical performance. The resonant frequency range of the sensors, with Circular Split-Ring Resonators CSRRs, is lower by 5% to 11% than the sensors with CSRRs. The directivity and gain of the sensors with CSRRs are higher by 2.5dB than the sensors without CSRRs. For S11 lower than –6 dB, the bandwidth of the novel metamaterial sensors may be around 15 to 55%. The directivity and gain of the new metamaterial sensors are around 5 dBi to 7.5 dBi. The receiving active sensor gain is 12 ± 3 dB. The transmitting active sensor gain is 13 ± 3 dB.

Keywords: wearable devices, wearable resonators, healthcare applications, IoT, metamaterial resonators, active antennas, communication, RF energy harvesting, self-powered devices

1. Introduction

Compact resonators and antennas are presented in [1–6]. The efficiency of compact resonators is low. Several types of compact wearable antennas, such as printed dipoles and microstrip antennas, are presented in [2–6]. Printed metamaterial sensors are employed in wireless communication systems and were presented in several publications, see [2–7]. Materials with periodic artificial structures are called metamaterials. The metamaterial structure and components define the electrical properties of the material. In [8–14] metallic posts structures and periodic split-ring resonators (SRRs) are used to produce structures with required permeability and dielectric constant. Metamaterials may be employed to develop efficient sensors for

RF, medical and IoT wearable devices [12–16]. The bandwidth and gain of the antenna presented in [8] are like those of patch antennas. Materials with negative dielectric permittivity are described in [9]. A model and setup to compute and measure the polarity of SRRs are described in [10]. A dual-band transmission-line metamaterial antenna with two transmission-line arms is described in [14]. The radiation efficiency of the antenna at 3.3 GHz is around 60% with 2.6 dBi directivity, and 0.8 dBi gain. Antennas such as patches, loops, and FIPA antennas have low efficiency [2–5, 16–29]. In communication and healthcare systems, the system polarization may be vertical, horizontal, or elliptical. In these systems, the radiating elements must be circular or dual-polarized. In [15–17] compact metamaterials sensors for healthcare devices are presented. Measurements of wearable sensors are presented in [19]. Active antennas for communication, medical, and IoT devices are presented in [20]. As presented in [30] Wearable healthcare devices are used to increase disease cure and prevention. A wireless device with thermal-aware protocol is presented in [30]. Wearable sensors and antennas for healthcare and RF systems are presented in [31–41]. Dual-polarized wearable antennas for healthcare applications are presented in [41]. Dual polarized metamaterial antennas have significant advantages over regular printed antennas, such as high efficiency and gain. The sensors electrical parameters on and near the human body were evaluated, see [2–6], by employing RF CAD software [42, 43].

In this chapter, metamaterials technology is used to develop high-efficiency sensors and antennas with harvesting energy units for medical, communication, and IoT devices. The energy harvesting units connected to the system provide compact self-powered efficient sensors.

The antennas bandwidth is around 20 to 45%, for VSWR, better than 3:1. The gain of the antennas with CSRRs is around 7.5 dB. The sensors efficiency is higher than 90%.

2. Wearable technologies and devices

Wearable systems and devices will be in the next following years, an important part of individuals' daily lives. Wearable healthcare systems may measure and record several medical parameters such as heartbeat rate, blood pressure, sweat rate, arterial blood pressure, body temperature, electrocardiograms, and electro-dermal activity. Wearable devices may provide monitoring and scanning features that are not provided by cellular phones and other computing devices. Wireless devices and technologies are employed to analyze and process the data collected by the healthcare devices. The collected medical information may be transmitted to the healthcare center to analyze the collected medical information. This process may send a message to the physician to call the patient that needs immediate medical treatment.

2.1 Features of wearable healthcare and sport devices

- Healthcare wearable systems monitor healthcare centers daily activities.
- Wearable systems monitor and operate companies' daily activities and systems.
- Healthcare wearable systems assist and help patients such as Asthma patients, Diabetes patients, Alzheimer's disease patients and Epilepsy patients.

- Healthcare wearable systems and devices may help to solve Obesity problems, Sleep disorders, and cardiovascular diseases.

Wearable healthcare systems may measure and record several medical parameters such as heartbeat rate, blood pressure, sweat rate, arterial blood pressure, body temperature, electrocardiograms, and electro-dermal activity. The application of WBANs in medical centers with energy harvesting units, where the medical parameters of a large number of patients are constantly being monitored, is presented in **Figure 1**.

2.2 IoT systems

IoT devices are wireless communication modules of interrelated computing systems, mechanical machines, personal devices, healthcare sensors, and digital machines that possess unique identifiers (UIDs). IoT devices and systems consist of wireless communication modules, processors, sensors, and radiating elements. IoT devices and systems transmit, receive, and process data from networks connected to the Internet web. IoT devices and systems are connected to an IoT gateway network where they gather data that is analyzed online or sent to information centers to be processed and shared with specific IoT modules and devices. The polarization of the receiving channel of IoT and medical devices should match the polarization of the transmitting channel. In several communication devices, the polarization is not defined. In these cases, the receiving antenna should be dual-polarized. IoT systems

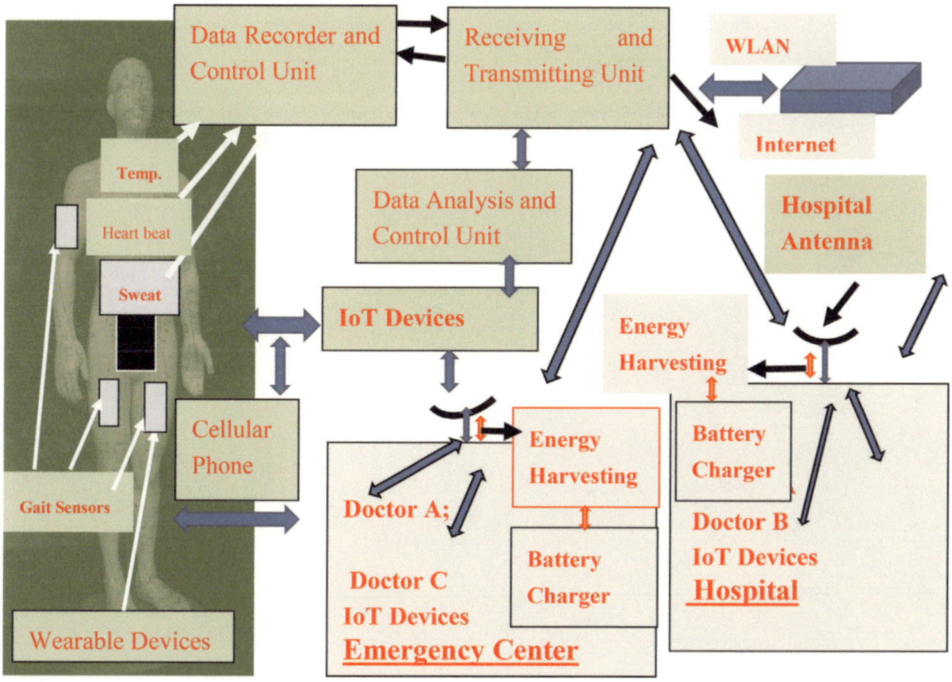

Figure 1.
Green wireless body area network, WWBAN, health monitoring sensors and system with energy harvesting units.

are significant in daily work and treatment in healthcare centers. IoT modules and systems may connect several healthcare devices and healthcare information centers to improve healthcare treatment. Efficient medical and IoT systems result in low-cost healthcare treatment.

- Efficient medical and IoT systems and devices are used to automate healthcare systems and procedures, to minimize organization and healthcare centers facilities and to lower labor costs.

- IoT systems may transmit data through a communication system without the need of human to computer interaction or human-to-human involvement.

- IoT systems can provide complete control over daily tasks and services in healthcare centers and organizations. IoT systems improve people's daily life and help companies to function more efficiently and smarter.

A block diagram of IoT medical device is presented in **Figure 2**.

IOT Devices and Systems Major Disadvantages-

- In several IoT devices and systems modules and information are shared between several machines and devices. This fact increases the threat that hackers can steal confidential data and corrupt devices connected to this IoT system.

- A bug in the IoT devices and system may corrupt devices and modules connected to this IoT system.

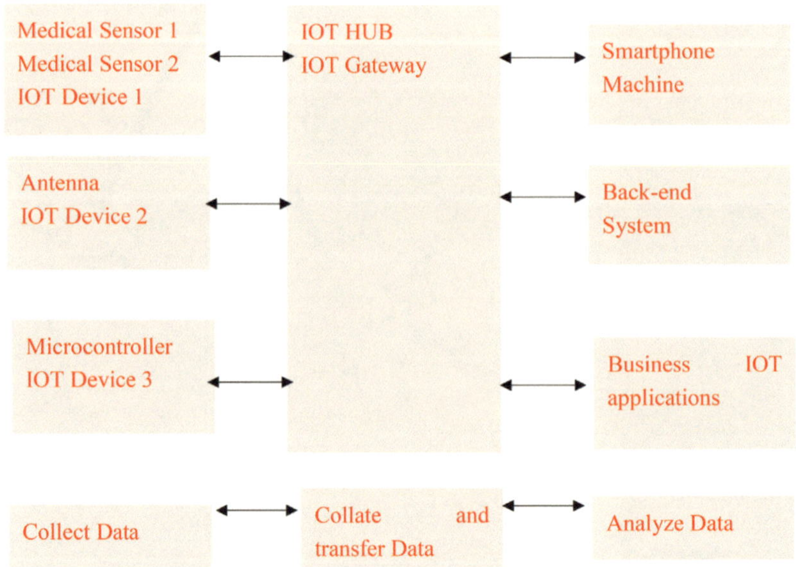

Figure 2.
IoT medical system block diagram.

2.3 Wireless communication devices and cellular phones features

The first generation of cellular phones began in the 1980s and used 1G technology. 1G technology of mobile phones uses analog radio systems. In the last decade, communication technology was improved rapidly and is employed in modern mobile phones. **Table 1** presents the improvement of mobile phone communication technologies.

5G Technology Main features

- High data transmission speed and high capacity.
- Connected and supportable to Wireless World Wide Web.
- Large broadcasting of data in Gbps.
- Observe TV programs with HD Clarity and read newspapers.
- Improved information transmission than that of 4G Generation.
- Better phone memory, improved dialing speed, improved clarity in audio and video features.
- 5G technology provides high resolution to cellular phone users and large bandwidth sharing.

3. Compact wearable dual polarized antenna for IoT and medical applications

The antenna with CSRR and metallic strips is presented in **Figure 3a**. The printed dipole matching network and the metallic strips are etched on the first layer with thickness of 0.16 cm. The radiating element with CSRR is printed on the second layer with thickness of 0.16 cm. The size of the antenna network with the energy harvesting module is 21.5×4.5 cm. The printed slot antenna is vertically polarized. The printed

Technology	1G	2G	3G	4G	5G
Data Speed	2.4 kbps	64 Kbps	3 Mbps	1Gbps	20 Gbps
Electronic Technology	Analog	Digital	Audio and video	Wi-Fi	Wi-Fi, wireless
Battery life	Few hours	Longer life cycle	Longer life cycle	Better life cycle	Power savings
Dimensions	Big and large	Smaller	Size reduction	Compact	Compact
Security	Low security	call and text encryption	Improved security	High security	High security
Capacity	Limited	Improved	Large	Large	High capacity

Table 1.
Comparison of cellular phone technologies.

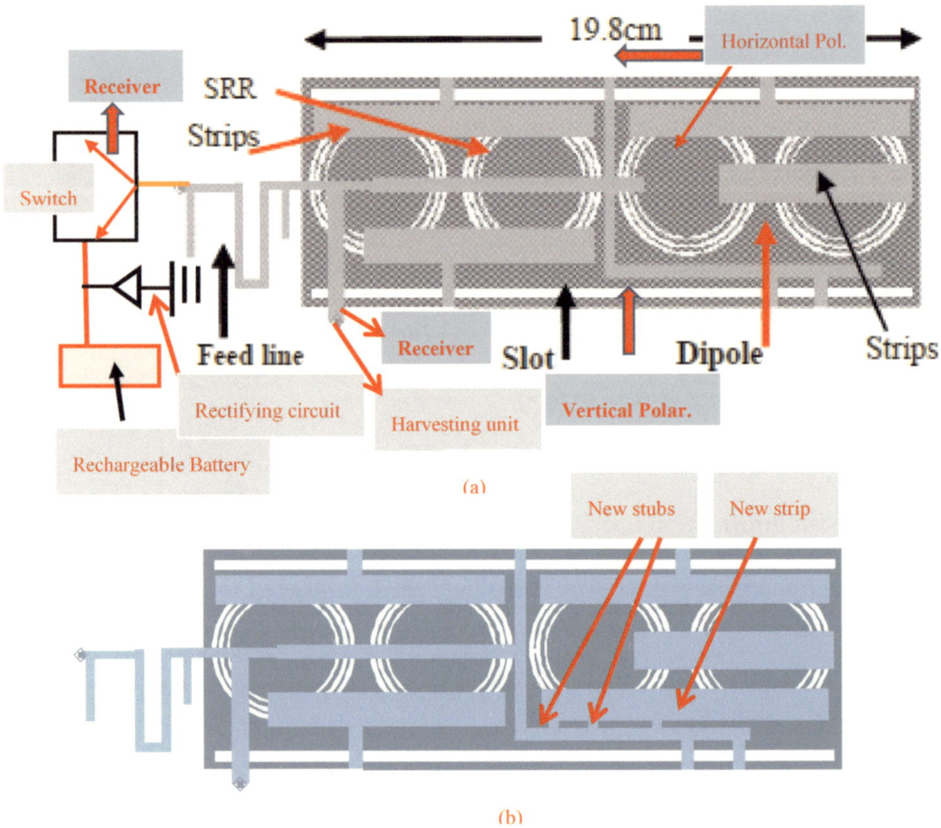

Figure 3.
(a). Antenna with CSRR and with energy harvesting unit, (b). optimized sensor with CSSR and metallic strips.

dipole is horizontally polarized. The resonant frequency of the dipole with CSRR is around 0.33GHz, which is lower by 15% than the resonant frequency of the printed dipole without CSRR. Many communications, IoT and healthcare systems operate in the frequency range between 0.1 and 0.55 GHz. The computed S11 and antenna gain are shown in **Figure 4**. The measured antenna bandwidth is around 45% for S_{11} lower

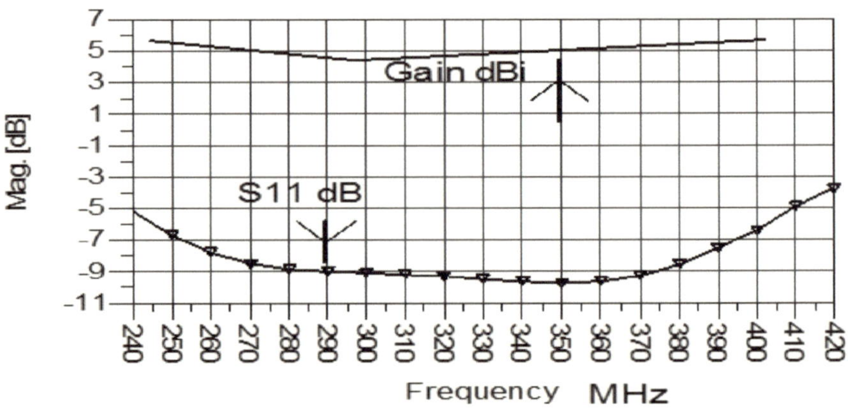

Figure 4.
Gain and S11 of the dual-polarized antenna with metallic strips and CSRR.

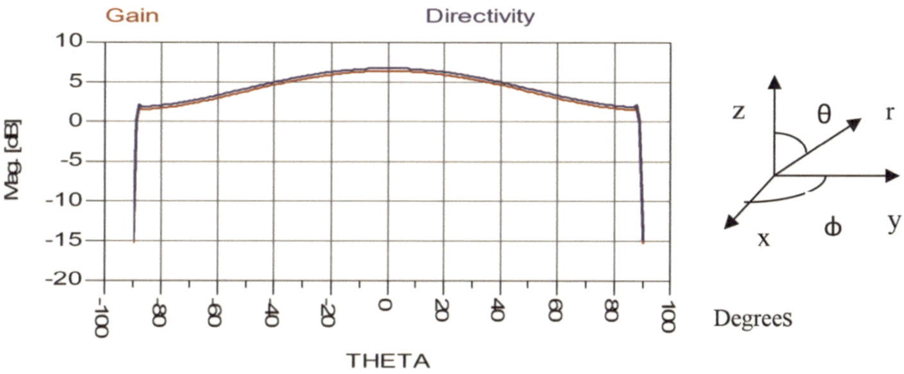

Figure 5.
Radiation pattern and gain of the antenna with metallic strips and CSRR.

than -6 dB. The dipole and the slot radiate in the z axis direction. The measured directivity and gain of the antenna with CSRR are around 5.5 dBi, as presented in **Figure 5**. The feed network of the antenna in **Figure 3a** was optimized, see **Figure 3b**, to yield S11 lower than -6 dB in the frequency range from 0.18 to 0.4 GHz, around 60% bandwidth, as presented in **Figure 6**. The dimensions of the antennas presented in this chapter are given in **Table 2**.

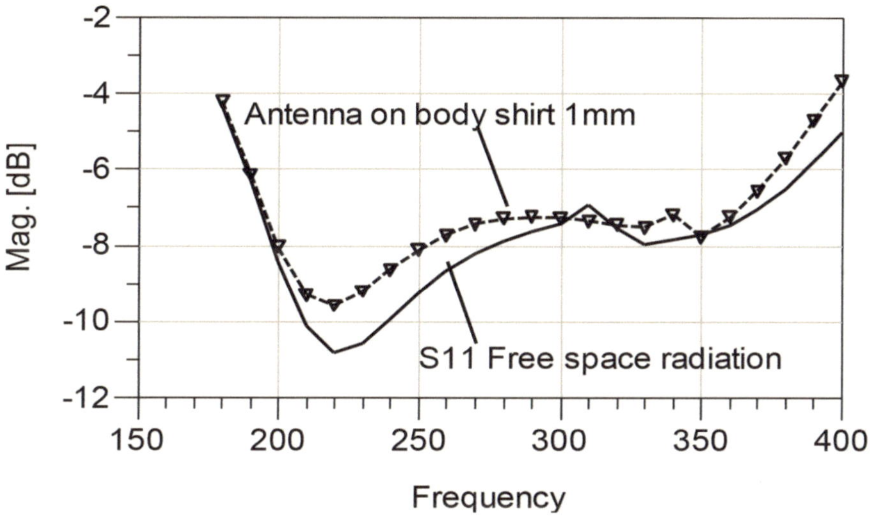

Figure 6.
S11 of the optimized dual-polarized antenna with metallic strips and CSRR on human body.

Antenna	Frequency (GHz)	BW %	Computed Gain dBi	Measured Gain dBi	Dimension (cm)	Efficiency %
Printed dipole with CSRR [2]	0.35	10	5.5	5.7	19.8×4.5	95
Dipole no CSRR [2]	0.4	10	2.5	2.5	21×4.5	90
Printed Loop	0.4–0.5	2	0	0	5 Diameter	Low
Active receiving loop	0.35–0.58	40	23 ± 2.5	22 ± 3	7 Diameter	50
Active transmitting loop	0.36–0.6	45	13 ± 3	12 ± 3	7 Diameter	50

Antenna	Frequency (GHz)	BW %	Computed Gain dBi	Measured Gain dBi	Dimension (cm)	Efficiency %
Stacked circular patch without CSRR [15]	2.7	8	5.5	5.3	4.8 Diameters	90
Circular patch with CSRR [15]	2.63	8	7.5	7.8	3.6 Diameters	85
Circular patch without CSRR [1, 2]	2.63	1.5	4.5	4.3	4.8 Diameters	85

Table 2.
Electrical performance comparison between wearable antennas without and with CSRR.

The sensors electrical parameters, presented in this chapter, on and near the human body were evaluated, see [2–7], by employing RF CAD software [42, 43]. The theoretical and equations used to design the sensors presented in this chapter are given in previous publications, see [2–7]. The energy harvesting module is connected to the antenna network feed line, as presented in **Figure 3**. The energy harvesting module can charge the battery when the switch is connected to the harvesting unit. Electromagnetic AC power is converted to DC power by employing a rectifying diode. The rectifier may be a half-wave rectifier or a full-wave rectifier. As shown in **Figure 3**, the harvesting module has an antenna, a rectifying circuit, and may recharge the device battery.

4. Wearable self-powered active sensor

A receiving self-powered active sensor is presented in **Figure 7**. The energy harvesting module acts as a dual-mode electromagnetic harvesting module. A switch may connect the LNA to the energy harvesting module. The energy harvesting module charges the battery. The Low Noise Amplifier is matched to the receiving antenna via a matching network. The LNA TAV541 is a linear PHMET amplifier. At 2 GHz, the amplifier has 0.45 dB Noise Figure and 18.5 dB gain. The LNA P1dB, at 2 GHz, is 19 dBm at the output port. The LNA specifications are listed in **Table 3**. A DC bias network supplies the required voltages to the amplifiers. The Low Noise Amplifier is matched to the receiver via an output-matching network. The sensor dimensions are around 21.5×5×1.9 cm. Gain and reflection coefficient of the metamaterial sensor is presented in **Figure 8**. The receiving active sensor gain as presented in **Figure 9** is 12 ± 3 dB, and the noise figure is better than 1 dB, for frequencies from 0.1GHz to 1GHz. The active metamaterial antenna was evaluated with Triquint LNA TQP3M9028. The amplifier specifications are given in **Table 3**. The active antenna gain with TQP3M9028 LNA is 11.1 ± 2.5 dB from 0.15 to 0.9GHz, as presented in **Figure 10**. The sensor noise figure, with TQP3M9028 LNA, for frequencies from 0.15GHz to 1GHz is better than 1.9 dB. The measured performance of the sensors with different LNAs is listed in **Table 4**. The active antennas with LNA TAV541 has better noise figure and higher gain. The sensor with LNA TQP3M9028 has a better 1dBc compression point and gain flatness.

5. Wearable self-powered active transmitting antenna

A transmitting self-powered active sensor is presented in **Figure 11**. The energy harvesting module acts as a dual-mode electromagnetic harvesting module.

DOI: http://dx.doi.org/10.5772/intechopen.112352

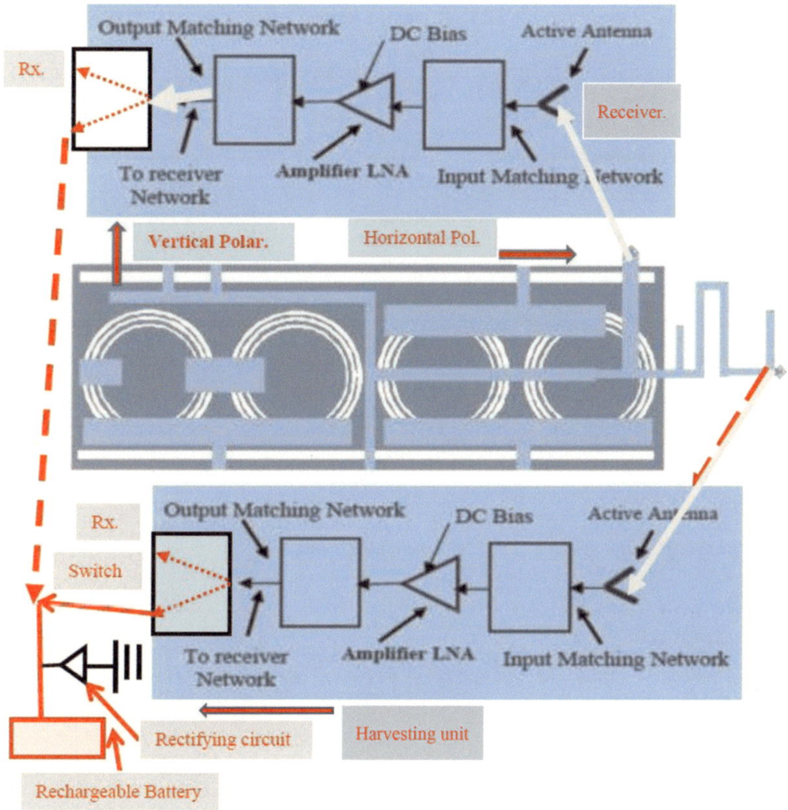

Figure 7.
Dual polarized receiving sensor with CSRR and with energy harvesting unit.

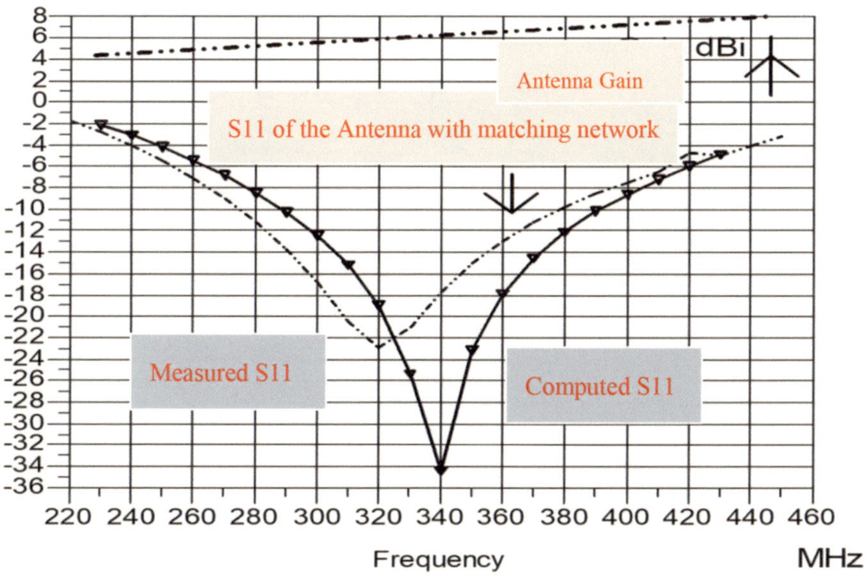

Figure 8.
S11 and gain of the dual-polarized antenna with CSRR and matching network.

Specification	TAV541, Mini Circuit	TQP3M9028, Triquint
Frequency, GHz	0.4–3	0.05–4
Gain at 2 GHz	18.5 dB	14.5 dB
N.F at 2GHz	0.45 dB	18 dB
P1dB at 2GHz	18.5 dBm	20.5 dBm
OIP3 at 2GHz	32 dBm	35 dBm
Operating Temp.°C	−40 – 80	−40 – 80
Max. Input power	17 dBm	17 dBm
Vgs, Volt	0.4	0.4
Vds	3 V, Ids = 60 mA	5 V, Ids = 85 mA
Supply voltage, V	±5	±5
Package type	Surface Mount	Surface Mount

Table 3.
Comparison of the specification of the S-band low noise amplifiers.

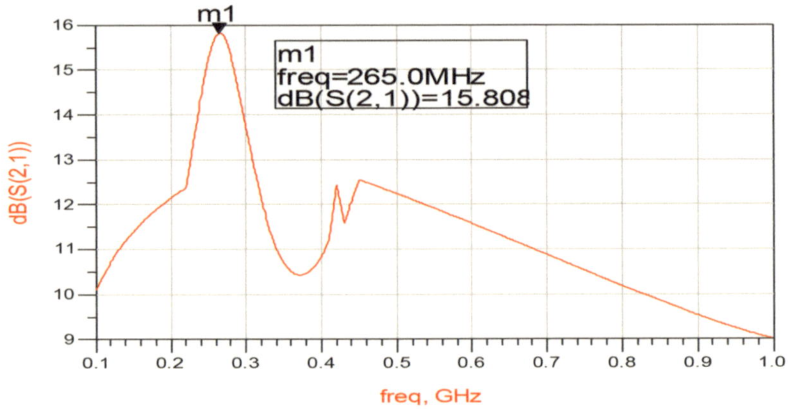

Figure 9.
Active receiving dual polarized receiving sensor gain, with LNA.

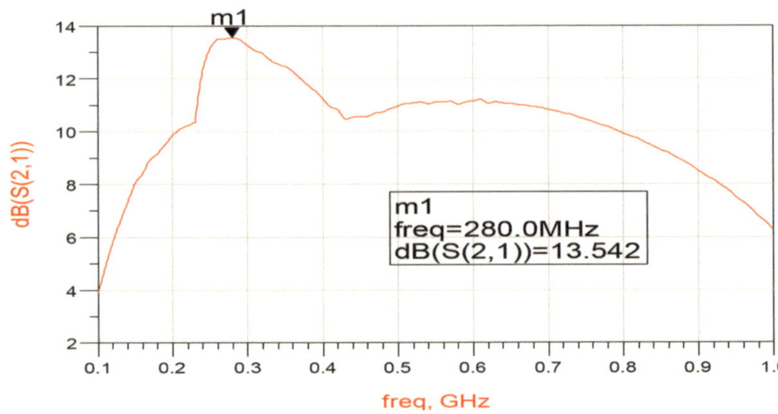

Figure 10.
Active receiving dual polarized receiving sensor gain, with TQP3M9028 LNA.

Parameter	Sensor with TAV541	Sensor with TQP3M9028
Frequency	0.1–1 GHz	0.15–0.9 GHz
VSWR	3:1	3:1
Gain	11 ± 2.5 dB	12 ± 3 dB
N.F at 1GHz	1 dB	2 dB
P1dB at 1GHz	19.0 dBm	20 dBm
Max Input power	17 dBm	17 dBm
Vgs, Volt	0.5	0.5
Vds	3 V, Ids = 55 mA	5 V, Ids = 80 mA
Dimensions	21.5×5×1.9 cm	21.5×5×1.9 cm

Table 4.
Comparison of the sensors measured performance with different LNA Amplifiers.

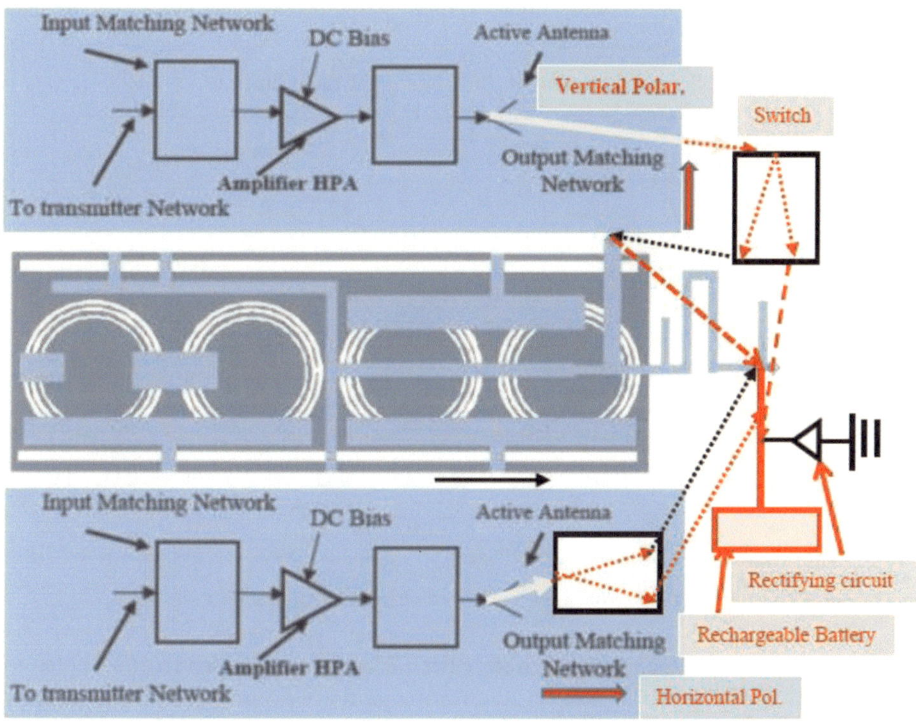

Figure 11.
Dual polarized transmitting sensor with CSRR and with energy harvesting unit.

The harvesting module can be part of a healthcare, IOT, and cellular phone. A switch may connect the transmitter battery to the energy harvesting module. The energy harvesting module charges the battery. Two power amplifiers were used to develop the metamaterial active sensor. The first amplifier is an MMIC GaAs MESFET VNA25, the second amplifier is an MMIC GaAs PHEMT HMC459. The amplifier specification is listed in **Table 5**. A DC bias network supplies the required voltages to the amplifiers. The sensor dimensions are around 21.5×5×1.9 cm. The active transmitting sensor

Specification	VNA25, Mini Circuit	HMC459, Triquint
Frequency GHz	0.4–2.5	DC–18
Gain dB at 1.8 GHz	17.5	16.5
N.F at 1.8 GHz	5.0 dB	3.8 dB
P1dB at 1.8 GHz	18.5 dBm	24 dBm
Max Input power	10 dBm	16 dBm
OIP3 at 1.8 GHz	28 dBm	29 dBm
Vgs, V	0.5	0.5
Vds	5 V, Ids = 85 mA	8 V, Ids = 290 mA
Supply voltage, V	±5	±8
Package	Surface Mount	Surface Mount
Temp. Range °C	−40 – 80	−40 – 80

Table 5.
Electrical Specification of the HPA Amplifiers.

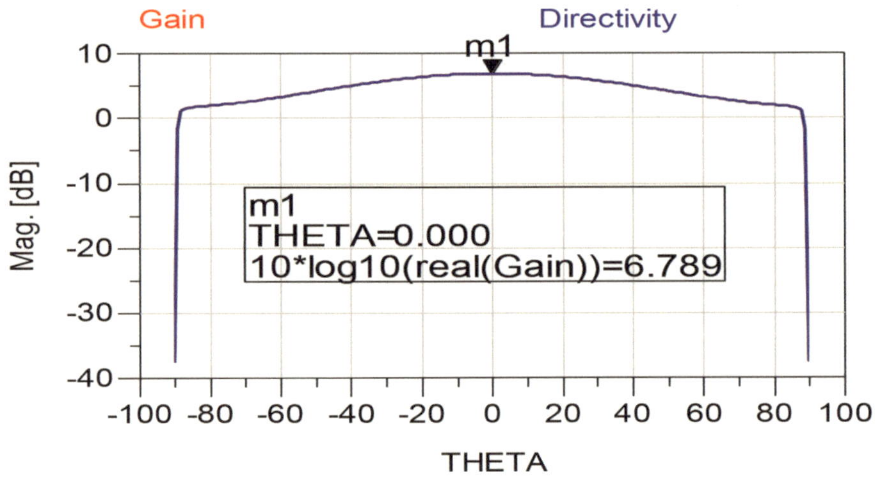

Figure 12.
Radiation pattern and gain of the dual-polarized antenna with metallic strips and CSRR.

VSWR, computed and measured, is better than 3:1 for frequencies from 0.25 to 0.45GHz. The computed and measured antenna gain are around 5.8 dBi, as presented in **Figure 12**. The active computed and measured antenna gain with HPA VNA25 is 13 ± 3 dB 1 for frequencies from 0.1 to 0.8 GHz. The transmitting sensor S21 parameter is shown in **Figure 13**. The active computed and measured antenna gain with the HMC459 HPA is 12 ± 4 dB for frequencies from 0.1 to 1 GHz, as shown in **Figure 14**. The transmitting sensor output power is around 19.5 dBm. The measured electrical performance of the sensors with the HPAs is listed in **Table 6**. The active antenna with VNA25 HPA has better gain flatness, higher gain, and lower DC power consumption. The transmitting antenna with HMC459 HPA has higher input and output power, and higher 1 dBC compression point. The HPA HMC459 has wider bandwidth from 1 to 18GHz. Photos of the metamaterial sensors, with CSSR and metallic strips, are shown in **Figure 15**.

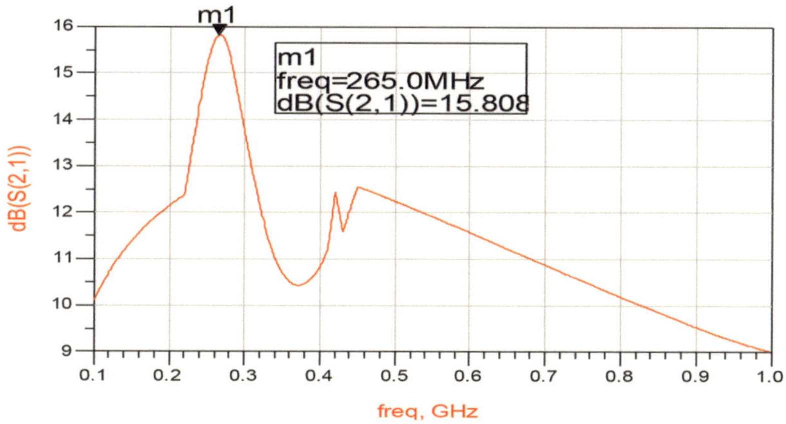

Figure 13.
Active transmitting dual polarized sensor gain, with HPA VNA25.

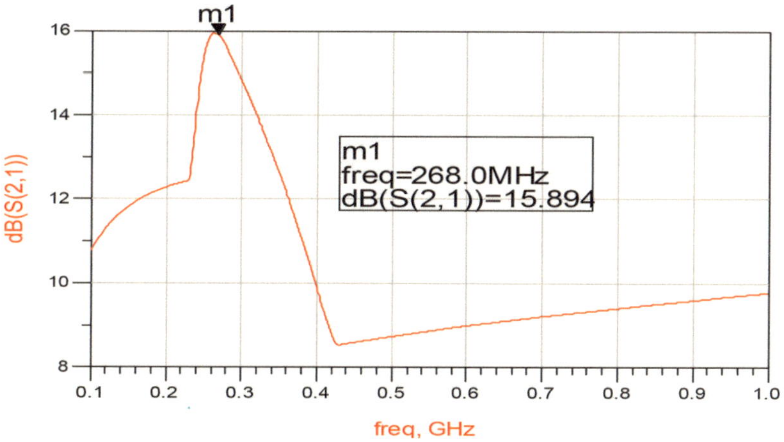

Figure 14.
Active transmitting dual polarized sensor gain, with HPA HMC459.

Parameter	Sensor with VNA25	Sensor with HMC459
Frequency	0.1–1GHz	0.15–0.9 GHz
VSWR	3:1	3:1
Gain	13 ± 3 dB	12 ± 4 dB
N.F at 1 GHz	6 dB	5 dB
P1dB at 1 GHz	19.0 dBm	24 dBm
Max Input power	10 dBm	16 dBm
Vgs	0.50 V	0.50 V
Vds	5 V, Ids = 85 mA	8 V, Ids = 290 mA
Dimensions	21×6×2cm	21×6×2cm

Table 6.
Comparison of the sensors performance with different HPA amplifiers.

(a)

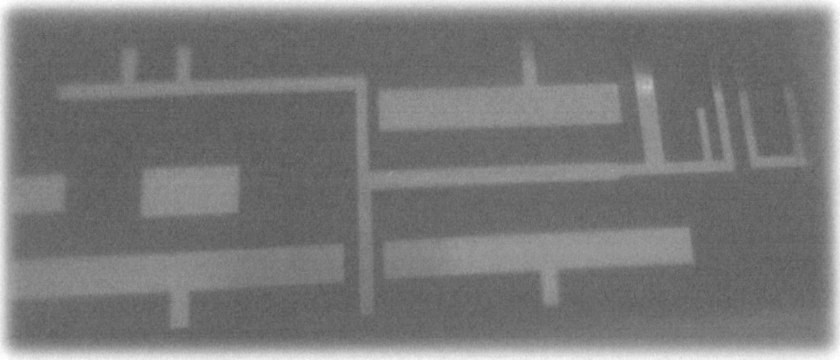

(b)

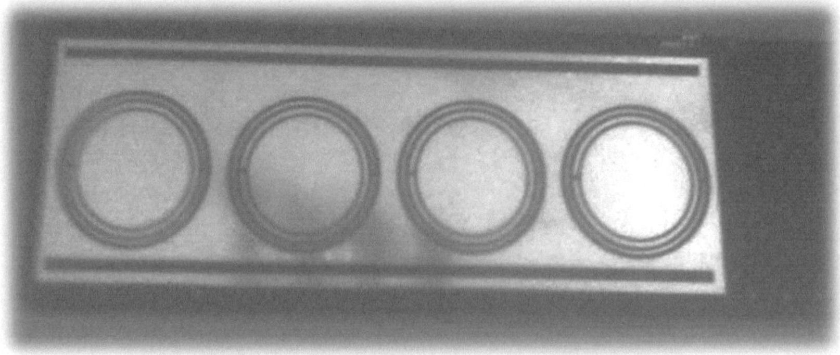

(c)

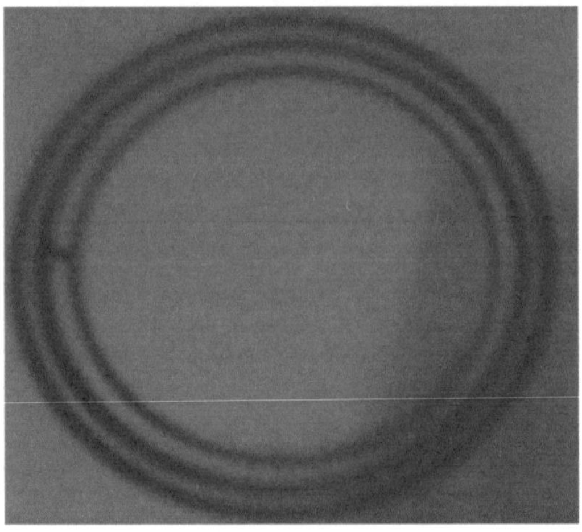

Figure 15.
Photos of the metamaterial antenna. a. Feed layer b. antenna with CSSRs c. CSSR.

6. Wearable metamaterial antennas for communication, healthcare, and IoT systems

The antenna's electrical parameters performance near the human body were computed by employing the antenna and human body model presented in **Figure 16a**. The dielectric constant and conductivity of human body tissues are given in **Table 7** [21, 22]. The effect of the antenna location on the human body is simulated by evaluating the antenna's electrical parameters on human body. The variation of the

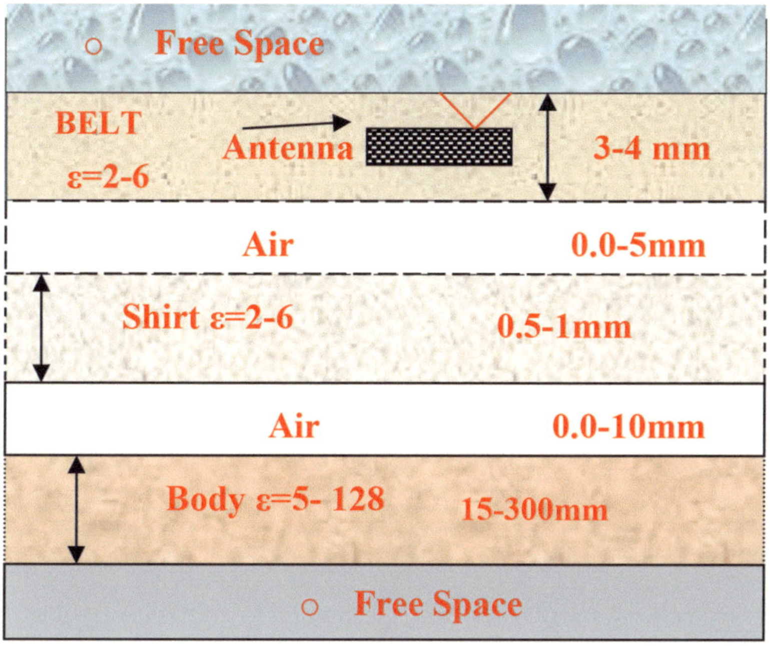

(a)

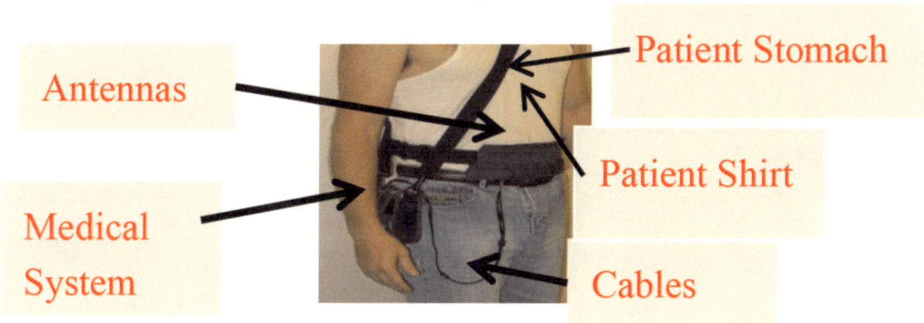

(b)

Figure 16.
(a) Model of wearable antenna environment; (b) wearable medical system on human body.

dielectric constant of the body tissues affects the electrical performance of the antenna. The antenna resonant frequency may be shifted up to 10% at different locations of the antenna on the patient's body. As shown in **Table 7**, the dielectric constant of fat tissues is 4.7, 41 in the skin tissues, 43 in the stomach area, and up to 128.1 in the small intestine tissues. The antennas may be placed inside a belt, as shown in **Figure 16b**. The antenna's electrical characteristics were computed and measured for air spacing between the sensors and human body up to 25 mm at different locations on the human body. Measurements of wearable sensors and antennas are done by using a phantom with sugar, salt, and water that represent the dielectric constant and conductivity of human body tissues [2–5, 19]. The antenna's electrical and mechanical characteristics were optimized to achieve the best antenna electrical and mechanical performance. Wearable antennas and sensors measurements and setup are discussed in [2–5, 19].

Tissue	Parameter	440 MHz	600 MHz	1 GHz	1.25 GHz
Fat tissues	σ	0.047	0.05	0.054	0.06
	ε	5.00	5.00	4.72	4.55
Stomach tissues	σ	0.71	0.75	0.96	0.98
	ε	42.7	41.40	39.66	39.00
Blood	σ	1.76	1.78	1.91	1.99
	ε	57.2	56.5	55.40	55.00
Skin	σ	0.58	0.6	0.63	0.77
	ε	41.6	40.45	40.25	39.65
Lung tissues	σ	0.27	0.27	0.27	0.28
	ε	38.4	38.4	38.4	38.4
Kidney tissues	σ	0.90	0.90	0.90	0.91
	ε	117.45	117.45	117.45	117.45
Colon tissues	σ	1.00	1.05	1.30	1.45
	ε	63.5	61.9	60.00	59.40
Small intestine	σ	1.74	1.74	1.74	1.74
	ε	128.1	128.1	128.1	128.1

Table 7.
Electrical parameters of human body tissues [16, 17].

Input Power dBm	Efficiency %	Remarks
−4−−6	8–12	Low Efficiency
−3−−2	28–32	Low Efficiency
−1 − +1	48–52	Good Efficiency
2–3	52–56	Good Efficiency
4–5	52–56	Good Efficiency
6–8	56–58	Good Efficiency
9–11	60–65	Best Efficiency

Table 8.
Measured harvester efficiency as function of input collected power.

Table 2 presents a comparison between computed and measured results of sensors without and with CSRR. **Table 8** presents a comparison of computed and measured results of compact wearable antennas.

Tables 2 and **8** verify that there is a good match between computed and measured results. Electrical performance of several passive and active antennas (such as dipoles, loop, slot and other antennas) is discussed in [2–6]. Medical, wearable wireless BAN, IoT, and monitoring systems with wearable BAN systems are shown in **Figure 17**. Hospitals, Health care centers and medical staff can be contacted from everywhere at any given time.

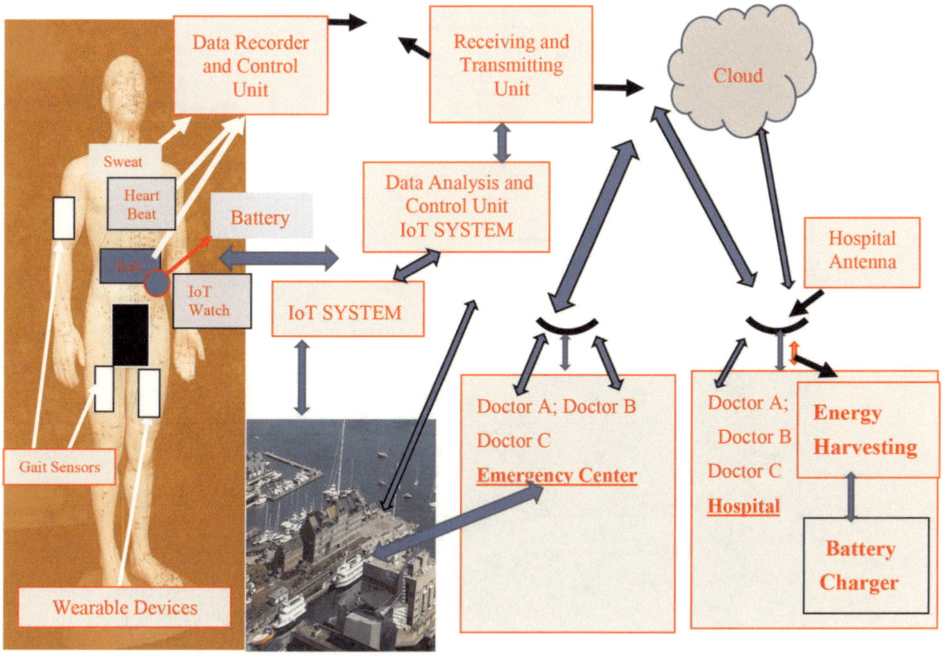

Figure 17.
Green medical, WWBAN, and IoT monitoring system with WBAN networks with energy harvesting units.

7. Electromagnetic energy harvesting concept

As presented in **Figure 3**, the energy harvesting unit consists of an antenna, rectifying diode and circuit, and a rechargeable battery. The energy harvesting unit and the radiating element provide a self-powered device. The rectifier diode converts electromagnetic energy, AC energy, to direct current (DC energy). Two types of diode rectifiers are usually used a half-wave rectifier or a full-wave rectifier [44–47]. A half-wave rectifier is presented in **Figure 18**. A half-wave rectifier converts only the half cycle of the positive voltage. It allows harvesting only one-half of the electromagnetic wave. The efficiency of the half-wave rectifier is around 41%. A full-wave bridge rectifier is shown in **Figure 19**. The bridge full-wave diode rectifier circuit converts electromagnetic energy to DC energy. The bridge rectifier consists of four diodes D1 through D4, as shown in **Figure 19**. The rectifier output DC voltage, $V_{ODC} = 2V_m/\pi$. By connecting a capacitor in shunt to the resistor, as shown in **Figure 19**, the rectifier output voltage may be improved. The full-wave rectifier

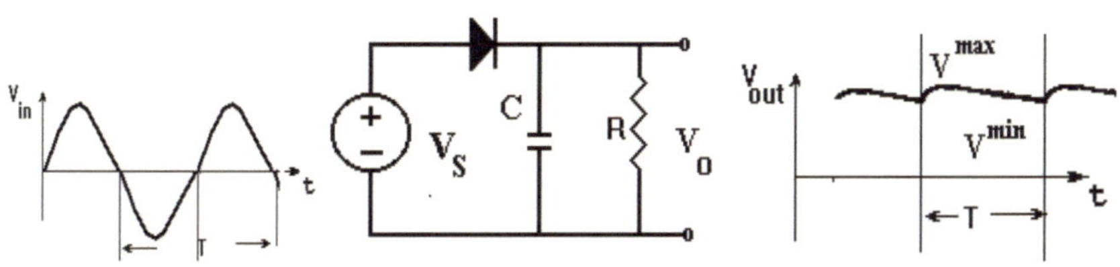

Figure 18.
Diode voltage rectifier with a capacitor, half wave.

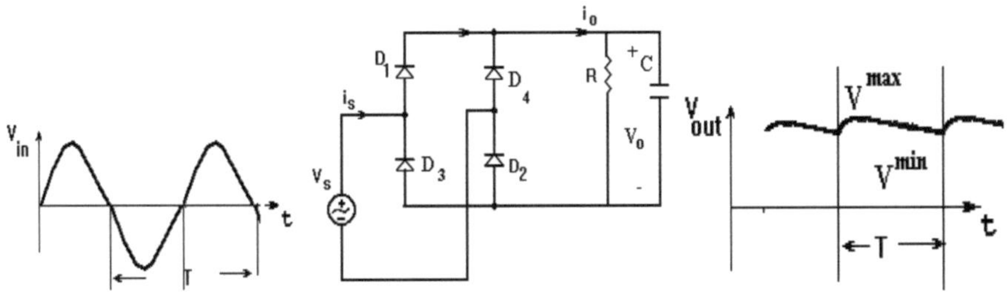

Figure 19.
Diode bridge voltage rectifier with a capacitor, full wave.

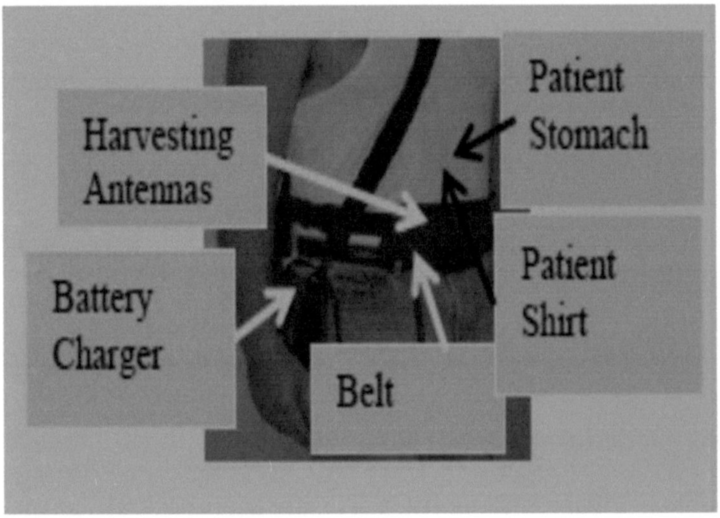

Figure 20.
Wearable RF system with energy harvesting unit for IoT, 5G, and healthcare systems.

efficiency is around 81%. Electromagnetic power amount in public centers, stadiums, hospitals, and malls may range from 1 $\mu W/cm^2$ to 5 mW/cm^2. The harvesting system efficiency increases as function of the electromagnetic power collected by the harvesting system, as listed in **Table 8**. The amplifier amplifies the input power collected by the energy harvesting system and improves the efficiency of the harvesting system. Results listed in **Table 8** are also presented by companies that manufacture commercial RF energy harvesting systems, see [2–7, 44–47]. If the RF radiating sources are close to the harvesting system, the RF power collected by the harvester will be higher.

Energy harvesting units provide green renewable energy and may eliminate the usage of power cords and the need to replace batteries frequently. Wearable medical devices with energy harvesting modules for wireless communication, IoT, and medical applications are presented in **Figure 20**. As presented in **Figure 20**, the medical device and harvesting module with a compact battery charger are located on the patient's shirt.

In 2022 almost everyone uses cellular phones, communication networks, tablets, and other RF communication systems. There is a huge increase in the amount of RF energy in the air. The expected amount of radio waves in the air in 2022 was around 60 Exa-bytes, EB, per month. The expected amount of RF energy in the air in 2025 is expected to be around 170 Exa-bytes per month. In RF energy harvesting modules, the radio frequency waves propagating in air can be received by the antennas and converted to DC power that is employed to charge batteries, sensors, and other wearable devices.

8. Conclusions

The active and passive sensors and metamaterials antennas presented in this chapter are wideband, efficient, compact, and low-cost. RF energy harvesting modules are connected to the antennas and sensors. Electromagnetic waves propagating in the air can be collected by the antennas and converted to DC power that may recharge the healthcare, computing system batteries, wearable sensors, and other RF modules. Evaluation of efficient active and passive wearable antennas and sensors is one of the most significant goals in the evaluation of wearable healthcare devices, medical sensors, IoT, wireless communication and medical systems. Passive and active metamaterial compact antennas and sensors characteristics such as gain, matching, noise figure, efficiency, bandwidth, and radiation pattern are presented in this chapter. The directivity and gain of the sensors with CSRRs are higher by 2.5 dB than the sensors without CSRRs. For S11 lower than -6 dB the bandwidth of the novel metamaterial sensors may be around 15 to 55%. The directivity and gain of the new metamaterial passive sensors are around 5 to 7.5 dBi. The receiving active sensor gain is 12 ± 3 dB. The transmitting active sensor gain is 13 ± 3 dB.

The metamaterial antennas and sensors discussed in this research may be used in wireless communication devices, computing networks, IoT devices, sport, and medical applications. Metamaterial technology is employed to design compact, efficient antennas and sensors. The receiving dual-polarized antenna network with the energy harvesting module, size is 21.5×4.5 cm. The printed slot antenna is vertically polarized. The printed dipole is horizontally polarized. The resonant frequency of the dipole with CSRR is around 0.33 GHz, which is lower by 15% than the resonant frequency of the printed dipole without CSRR. The measured antenna bandwidth is around 45% for S11 lower than -6 dB. The measured directivity and gain of the antenna with CSRR are around 5.5 dBi. The S11 of the optimized receiving metamaterial antenna is lower than -6 dB in the frequency range from 0.18 to 0.4 GHz, around 60% bandwidth. The receiving active sensor gain with TAV541 LNA is 12 ± 3 dB, and the noise figure is better than 1 dB, for frequencies from 0.1 to 1GHz.

The active computed and measured transmitting antenna gain with HPA VNA25, is 13 ± 3 dB 1 for frequencies from 0.1 to 0.8 GHz. The transmitting sensor output power is around 19.5 dBm.

Electromagnetic power amount in public centers, stadiums, hospitals, and malls may range from 1 $\mu W/cm^2$ to 5 mW/cm^2. The harvesting system efficiency increases as function of the RF power collected by the harvesting system. The efficiency of the harvesting system is around 50% for 0 dBm input power. However, the efficiency of the harvesting system is around 60% for 10 dBm input power.

The antennas and sensors presented in this chapter may be used in medical devices that improve the daily health of patients. Wearable antennas and healthcare devices are an important choice for healthcare organizations, hospitals, and patients. The wearable sensors presented in this chapter support the development of personal healthcare systems with online immediate medical staff responses to treat and improve patients' health.

Author details

Albert Sabban[1,2]

1 Department of Electrical Engineering, Ort Braude College, Karmiel, Israel

2 Senior RF and communication systems researcher and consultant in High Tech Companies, Haifa, Israel

*Address all correspondence to: sabban@netvision.net.il

References

[1] James JR, Hall PS, Wood C. Microstrip Antenna Theory and Design. London, UK: Institution of Engineering and Technology (IET); 1981

[2] Sabban A. New compact wearable metamaterials circular patch antennas for IoT, medical and 5G applications, MPDI. Applied System Innovation. 2020;**3**(4):42. DOI: 10.3390/asi3040042

[3] Sabban A. Wearable Communication Systems and Antennas. 2nd ed. Bristol, UK: IOP Publishing Ltd; 2022. ISBN: 978-0-7503-5220-8

[4] Sabban A. Wearable Circular Polarized Antennas for Health Care, 5G, Energy Harvesting, and IoT Systems. Electronics. MDPI Publication. Jan 2022. (ISSN 2079-9292)

[5] Sabban A. Novel Wearable Antennas for Communication and Medical Systems. Boca Raton, FL, USA: Taylor & Francis Group; 2017

[6] Sabban A. Wideband RF Technologies and Antennas in Microwave Frequencies. Hoboken, NJ, USA: Wiley; 2016

[7] Sabban A. Low Visibility Antennas for Communication Systems. Boca Raton, FL, USA: Taylor & Francis; 2015

[8] Rajab K, Mittra R, Lanagan M. Size reduction of microstrip antennas using metamaterials. In: Proceedings of the 2005 IEEE APS Symposium. Washington, DC, USA: IEEE; 3–8 Jul 2005

[9] Pendry JB, Holden AJ, Stewart WJ, Youngs I. Extremely low frequency Plasmons in metallic Meso-structures. Physical Review Letters. 1996;**76**: 4773-4776. DOI: 10.1103/physrevlett. 76.4773

[10] Pendry JB, Holden A, Robbins D, Stewart W. Magnetism from conductors and enhanced nonlinear phenomena. IEEE Transactions on Microwave Theory and Techniques. 1999;**47**:2075-2084. DOI: 10.1109/22.798002

[11] Marqués R, Mesa F, Martel J, Medina F. Comparative analysis of edge and broadside coupled split ring resonators for metamaterial design—Theory and experiment. IEEE Transactions on Antennas and Propagation. 2003;**51**:2572-2581

[12] Marqués R, Baena JD, Martel J, Medina F, Falcone F, Sorolla M, et al. Novel small resonant electromagnetic particles for metamaterial and filter design. In: Proceedings of the ICEAA'03, Symposiums. Torino, Italy. 8–12 September 2003. pp. 439-442

[13] Marqués R, Martel J, Mesa F, Medina F. Left-handed-media simulation and transmission of EM waves in subwavelength Split-ring-resonator-loaded metallic waveguides. Physical Review Letters. 2002;**89**:183901. DOI: 10.1103/ physrevlett.89.183901

[14] Zhu J, Eleftheriades G. A compact transmission-line metamaterial antenna with extended bandwidth. IEEE Antennas and Wireless Propagation Letters. 2008;**8**:295-298. DOI: 10.1109/ LAWP.2008.2010722

[15] Baena JD, Marqués R, Martel J, Medina F. Experimental results on metamaterial simulation using SRR-loaded waveguides. In: Proceedings of the IEEE- AP/S International

Symposium on Antennas and Propagation, Columbus, OH, USA. 22–27 June 2003. pp. 106-109

[16] Marques R, Martel J, Mesa F, Medina F. A new 2D isotropic left-handed metamaterial design: Theory and experiment. Microwave and Optical Technology Letters. 2002;**35**:405-408. DOI: 10.1002/mop.10620

[17] Sabban A. Small wearable metamaterials antennas for medical systems. Applied Computational Electromagnetics Society Journal. 2016; **31**:434-443

[18] Sabban A. Microstrip antenna arrays. In: Nasimuddin N editor. Microstrip Antennas. London, UK: InTech; November 2011. pp. 361–384. ISBN 978-953-307-247-0. Available from: http://www.intechopen.com/articles/show/title/microstripantenna-arrays

[19] Sabban A. Wearable antenna measurements in vicinity of human body. Wireless Engineering and Technology. 2016;7:97-104. DOI: 10.4236/wet.2016.73010

[20] Sabban A. Active compact wearable body area networks for wireless communication, medical and IOT applications. MDPI ASI, Applied System Innovation Journal. 2018;**1**(4):46. DOI: 10.3390/asi1040046

[21] Chirwa L, Hammond P, Roy S, Cumming DRS. Electromagnetic radiation from ingested sources in the human intestine between 150 MHz and 1.2 GHz. IEEE Transactions on Biomedical Engineering. 2003;**50**: 484-492. DOI: 10.1109/tbme.2003.809474

[22] Werber D, Schwentner A, Biebl EM. Investigation of RF transmission properties of human tissues. Advances in Radio Science. 2006;**4**:357-360. DOI: 10.5194/ars-4-357-2006

[23] Gupta B, Sankaralingam S, Dhar S. Development of wearable and implantable antennas in the last decade: A review. In: Proceedings of the 2010 10th Mediterranean Microwave Symposium, Guzelyurt, Cyprus. 25–27 August 2010. pp. 251-267

[24] Thalmann T, Popovic Z, Notaros BM, Mosig JR. Investigation and design of a multi-band wearable antenna. In: Proceedings of the 3rd European Conference on Antennas and Propagation, EuCAP 2009, Berlin, Germany. 23–27 March 2009. pp. 462-465

[25] Salonen P, Rahmat-Samii Y, Kivikoski M. Wearable antennas in the vicinity of human body. In: Proceedings of the IEEE Antennas and Propagation Society Symposium, Monterey, CA. 20–25 June 2004. pp. 467-470

[26] Kellomäki T, Heikkinen J, Kivikoski M. Wearable antennas for FM reception. In: Proceedings of the 2006 First European Conference on Antennas and Propagation, Nice, France, 6–10 November. 2006. pp. 1-6

[27] Lee Y. Antenna Circuit Design for RFID Applications; Microchip (Application Note 710c). Chandler, AZ, USA: Microchip Technology Inc.; 2003

[28] Sabban A. Microstrip Antenna and Antenna Arrays. U.S. Patent US 4,623,893. 18 November 1986

[29] Wheeler HA. Small antennas. IEEE Transactions on Antennas and Propagation. 1975;**23**:462-469

[30] Jamil F, Ahmad S, Iqbal N, Kim D. Towards a remote monitoring of patient

vital signs based on IoT-based Blockchain integrity management platforms in smart hospitals. Sensors. 2020;**20**:2195. DOI: 10.3390/s20082195

[31] Jamil F, Iqbal MA, Amin R, Kim D. Adaptive thermal-aware routing protocol for wireless body area network. Electronics. 2019;**8**:47. DOI: 10.3390/electronics8010047

[32] Shahbazi Z, Byun Y-C. Towards a secure thermal-energy aware routing protocol in wireless body area network based on Blockchain technology. Sensors. 2020;**20**:3604. DOI: 10.3390/s20123604

[33] Lin J, Itoh T. Active integrated antennas. IEEE Transactions on Microwave Theory and Techniques. 1994;**42**:2186-2194. DOI: 10.1109/22.339741

[34] Mortazwi A, Itoh T, Harvey J. Active Antennas and Quasi-Optical Arrays. New York, NY, USA: John Wiley & Sons; 1998

[35] Jacobsen S, Klemetsen Ø. Improved detectability in medical microwave radio-thermometers as obtained by active antennas. IEEE Transactions on Biomedical Engineering. 2008;**55**:2778-2785. DOI: 10.1109/tbme.2008.2002156

[36] Jacobsen S, Klemetsen O. Active antennas in medical microwave radiometry. Electronics Letters. 2007;**43**:606. DOI: 10.1049/el:20070577

[37] Ellingson SW, Simonetti JH, Patterson CD. Design and evaluation of an active antenna for a 29–47 MHz radio telescope array. IEEE Transactions on Antennas and Propagation. 2007;**55**:826-831

[38] Segovia-Vargas D, Castro-Galan D, Munoz LEG, González-Posadas V. Broadband active receiving patch with resistive equalization. IEEE Transactions on Microwave Theory and Techniques. 2008;**56**:56-64. DOI: 10.1109/TMTT.2007.912008

[39] Rizzoli V, Costanzo A, Spadoni P. Computer-aided Design of Ultra-Wideband Active Antennas by means of a new figure of merit. IEEE Microwave a nd Wireless Components Letters. 2008;**18**:290-292. DOI: 10.1109/LMWC.2008.918971

[40] Sabban A. A new wideband stacked microstrip antenna. In: Proceedings of the IEEE Antenna and Propagation Symposium. Houston, TX, USA: IEEE; May 1983. pp. 23-26

[41] Sabban A. Dual Polarized Dipole Wearable Antenna. U.S Patent 8203497. 19 June 2012

[42] ADS Momentum Software, Keysightt. Available from: http://www.keysight.com/en/pc-1297113/advanced-design-system- ads?cc=IL&lc=eng [accessed on 3 January 2018]

[43] Sabban A. New wideband printed antennas for medical applications. IEEE Transactions on Antennas and Propagation. 2013;**2013**(61):84-91. DOI: 10.1109/tap.2012.2214993., January

[44] Paradiso JA, Starner T. Energy scavenging for mobile and wireless electronics. IEEE Pervasive Computing. 2005;**4**(1):18-27

[45] Valenta CR, Durgin GD. Harvesting wireless power: Survey of energy-harvester conversion efficiency in far-field, wireless power transfer systems. IEEE Microwave Magazine. 2014;**15**(4):108-120

[46] Nintanavongsa P, Muncuk U, Lewis DR, Chowdhury KR. Design

optimization and implementation for RF energy harvesting circuits. IEEE Journal on Emerging and Selected Topics in Circuits and Systems. 2012; **2**(1):24-33

[47] Devi KKA, Sadasivam S, Din NM, Chakrabarthy CK. Design of a 377 Ω patch antenna for ambient RF energy harvesting at downlink frequency of GSM 900. In: Proceedings of the 17th Asia Pacific Conference on Communications (APCC'11), Sabah, Malaysia. October 2011. pp. 492-495

Chapter 3

Variable Renewable Energy: How the Energy Markets Rules Could Improve Electrical System Reliability

Daniel Llarens, Laura Souilla, Santiago A. Masiriz and Gastón R. Lestard

Abstract

In the last 10 years, significant changes have been observed in the operation of electrical systems resulting from the increasing incorporation of Variable Renewable Energy (NCRE—Solar PV, WIND) characterized by strong volatility in its energy production, due to climatic effects, which affect the reliability in the operation of the electrical system. These technologies also show a significant reduction in their capital costs, which are currently competitive compared to conventional alternatives for energy production, with the advantage of contributing to reducing the production of greenhouse gases. Therefore, increasing reliability operational problems are expected in the future, which must be resolved to supply the demand safely and at minimum cost. LATAM's countries are making slow progress in updating their regulatory frameworks for the electricity sector to include changes that improve the integration of NCRE generation without reducing the quality of service. This document describes possible regulatory changes that could be implemented to promote a system safe operation including (a) intra-hours marginal costs, (b) day-ahead/intraday energy markets, (c) incentives to better forecast the NCRE generation production profile, (d) participation of NCRE generation in the capacity market, and (e) including BESS as ancillary service for frequency/ramp power control.

Keywords: electricity markets, reserves, economic signals, marginal costs, reliability, firm capacity, ancillary services, distribution rates

1. Introduction

In the LATAM's countries, mainly during the 1990s, wholesale electricity markets were created as a means to ensure that the restructuring process of the electricity sector leads to private participation in the efficient development of the sector. The legislation and regulation of the markets establish economic signals for investment and operation decisions given by energy prices that seek to reflect the marginal cost of the system and other economical signals to promote the efficient expansion of the generation fleet.

The regulatory framework of each market (law and other regulations) establishes the objectives, criteria, and general rules that define the framework in which the market must be developed and the types of agents that can participate commercially. Likewise, they define the tariff criteria and update methodologies that result in the tariffs that are applied to energy transactions and transmission charges.

Generators have free access to the transmission network and thus to the spot market where they sell their energy production at prices that arise from competition for dispatch based on the variable production cost of each power plant. Market prices represent the short-run marginal cost of the market.

Within this conceptual framework, LATAM's electricity markets have shown great dynamism since their inception with strong private participation during the privatization processes of state-owned companies and in new generation investment projects.

The generation of electrical energy in these markets was traditionally carried out based on conventional generation (thermal, hydraulic, geothermal, and nuclear) characterized by production patterns known in advance, as a result of which the reliability of the electrical supply was high, and the events that affected the quality of electrical service limited to the unavailability of some generator/element of the transmission system due to forced unavailability resulting from an unanticipated failure.

In the last 10 years, significant changes have been observed in the operation of these electrical systems resulting from the increasing incorporation of nonconventional renewable generation NCRE (mainly solar PV and WIND) characterized by strong volatility in its energy production, due to climatic effects, which affect the reliability in the operation of the electrical system [1–5].

These technologies also show a significant reduction in their capital costs [6], which are currently competitive compared to conventional alternatives for energy production, with the advantage of contributing to reducing the production of greenhouse gases.

The electricity production based on NCRE is currently a key element in the energy transition that the electricity sectors must endure, in the following decades, toward an operation with low greenhouse gas emissions.

However, the penetration of NCRE technologies, given precisely their variability, can alter the functioning of the electricity markets. The particular characteristics of NCRE, considered non-manageable, produce technical-economic impacts on the operation of the system.

This situation, which should not prevent the development of these technologies to take advantage of their advantages, however, requires the introduction of reforms in the functioning of the market to efficiently assimilate the high levels of NCRE penetration that are forecast for the near future.

As NCRE penetration levels grow, many electricity sectors around the world are already facing new challenges and respective reforms are taking place to improve the integration of these technologies into the market.

Although the impact of renewable technologies varies according to the characteristics of the system and the regulatory environment, the main problems that arise are similar.

In recent years, a significant reduction in the cost of storage media, mainly batteries (BESS), has also been observed. The storage facilities could contribute to mitigating the volatility of the production of renewables by providing quickly managed reserves. The reduction in the development costs of BESS also makes it possible to anticipate their rapid growth, helping to mitigate the operational problems of the electrical system.

In systems with variable renewable generation, the system reserves must compensate for random variations in demand and generation. The demand has daily cycles. Variable renewable generators add variation patterns to their production. Tracking variability may require more operational flexibility and more frequent dispatch adjustments.

In the particular case of LATAM's countries, given the growing participation of NCRE generation, restrictions have been introduced that tend to limit the participation of this type of technology in the market as a way of guaranteeing security in the supply of demand.

Some countries are making slow progress in updating their regulatory frameworks for the electricity sector to include changes that improve the integration of NCRE generation without reducing the quality of service.

This document describes possible regulatory changes that could be implemented to promote a system safe operation at minimum cost with high penetration of ERNC generation. It includes (a) intra-hours marginal costs, (b) day-ahead/intraday energy markets, (c) incentives to better forecast the NCRE generation production profile, (d) participation of NCRE generation in the capacity market, and (e) including BESS as an ancillary service for frequency/ramp power control.

The document is organized into three sections, which are as follows:

Section #1: Presents a general description of the operation of the LATAMs electricity markets. It includes how the generation fleet is dispatched and how is determined the marginal cost of energy, the composition of the generation fleet, and the growth of ERNC generation.

Section #2: Presents some observed operational problems that affect the security of supply resulting from the volatility in the production of wind and solar generation.

Section #3: Describe market rules that could mitigate the adverse effects of intermittent generation, favoring its participation in the electricity markets without affecting the reliability of the electrical system. It includes: (a) intra-hours marginal costs, (b) day-ahead/intraday energy markets, (c) optimization of system reserves, (d) incentives to better forecast the NCRE generation production profile, (e) participation of NCRE generation in the capacity market, (f) including BESS as an ancillary service for frequency/ramp power control, and (g) demand participation for increasing the system reserves.

2. LATAM's market operation

2.1 Generation dispatch

Each of LATAM's electricity markets has particular characteristics concerning the mechanism used to perform the generation dispatch. Even so, there are common criteria that are summarized below.

The generation fleet is dispatched based on the variable production cost (VPC) of each generator. The VPC is equal to the sum of variable fuel cost plus O&M cost.

$$\text{VPC}\left[\frac{\text{USD}}{\text{MWh}}\right] = \text{Eficiency} \times \text{Fuel Cost} + \text{O\&M} \quad (1)$$

The generator with the lowest VPC is dispatched first (typically renewable generators (hydro, solar, and wind) followed by clean thermal generation (geothermal, efficient cogeneration, and nuclear) and lastly conventional thermal generation based on coal, natural gas, and liquid fuels (fuel oil and diesel). The generation dispatch order is called LIST OF MERIT.

The LIST OF MERIT may be changed due to technical requirements associated with the security of supply (reliability constraints).

Figure 1 shows (on the right) a simple example of building the LIST OF MERIT following the criteria indicated above. Each rectangle represents one particular power plant. The height is the available power of the plant, and the width is 1 hour, so the rectangle area represents the available energy of the power plant in each hour.

The generators with the lowest VPC, typically renewables (in green) are located first followed by hydro generation (in blue), and lastly, the thermal generators (T1–T5 in brown) that are ordered by their variable production costs (VPC) the cheapest first.

Figure 1 shows (on the left) the hourly demand of the system. Typically, the demand is minimum in the early morning hours and maximum in the afternoon/night hours.

The demand for 1 hour in particular (hour 23, red dot) is supplied by the generators with the lower VPC of the LIST OF MERIT: renewable generation plus hydro generation and thermal power plants T1, T2, T3, and partially T4. The T5 generator is not dispatched because it is the one with the highest VPC.

When demand is reduced, the dispatch of thermal generation is reduced. In the figure, at hour 7, only generators T1 and T2 (partial) are dispatched.

The marginal cost of generation in each hour is equal to the VPC of the generator with the highest VPC that is dispatched according to the aforementioned methodology. In the example, the marginal cost at hour 23 is equal to the VPC of generator G4, and the marginal cost at hour 7 is equal to the CVP of generator G2.

Therefore, the marginal costs are higher when the demand is higher (and therefore the lower the reserve margin), resulting in the marginal cost being an economic signal that promotes the availability of generation as a way of achieving maximum profitability of the generation business.

Figure 2 shows a typical generation dispatch by type. The figure shows the generation dispatch for each hour, for a typical day of each month (12 × 24 matrix) in the Mexican Electricity Market (MX).

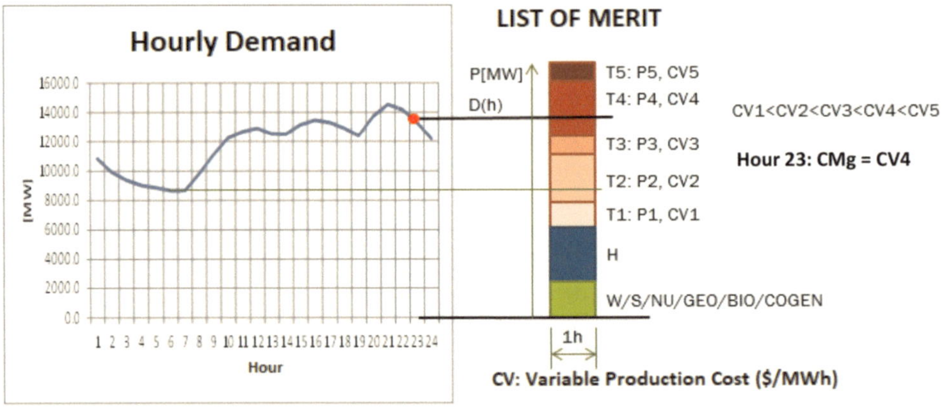

Figure 1.
Generation dispatch methodology (LIST OF MERIT).

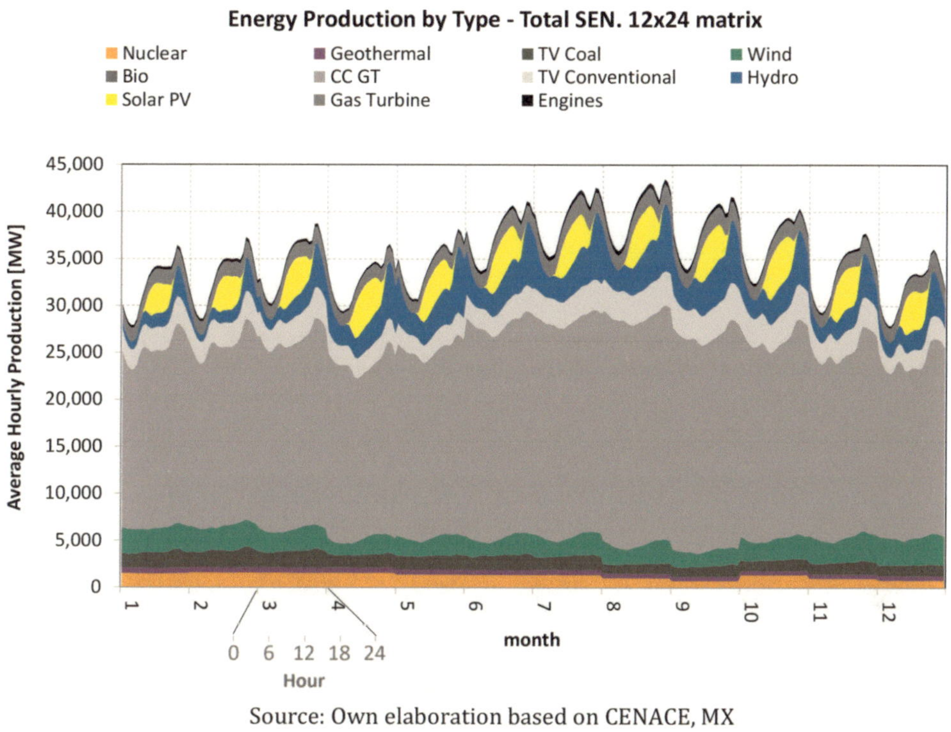

Figure 2.
Energy production by type. 12 × 24 matrix. Source: Own elaboration based on CENACE, MX.

Hydro generation is maximum at night, which compensates for the zero solar generation in those hours. Thermal generators with the lowest VPC are dispatched as baseload (typically geothermal, nuclear, combine cycle running with natural gas and coal thermal plants). High expensive thermal plants (conventional thermal running with liquid fuels) are dispatched last, as peaking units.

Generation dispatch includes the system support resources (RSS).[1] They are generators that are dispatched to maintain the reliability of the power system.

2.2 Addition of new generation capacity

In the LATAM's electricity markets, the expansion of the generation fleet that arises from the initiative of private investors is mainly based on natural gas combined cycle type thermal plants (distributed by networks or LNG type) and WIND and SOLAR PV renewable plants.

The main reason for the development of those kinds of projects is because their development cost (CAPEX + OPEX) is minimal and they allow compliance with renewable generation participation quotas imposed by governments.

As an example, **Figure 3** shows the installed generation capacity in MX by type, period 2017 to 2020. In this period, the ERNC generation increases by 9072 MW and CCGT generation by 11,490 MW. The other technologies show minimum changes (**Table 1**).

[1] For example, generators required to run out-of-merit to provide security or reliability in a given area.

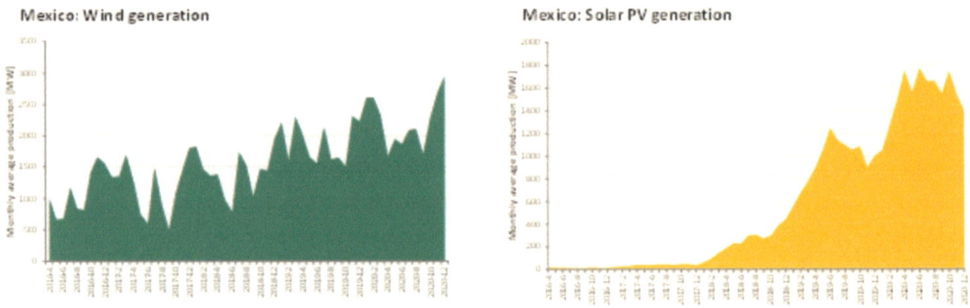

Figure 3.
Production of WIND/SOLAR PV generation. Source: Own elaboration based on CENACE, MX.

Technology	2017	2018	2019	2020
Hydro	12612	12612	12612	12614
Geothermal	899	899	899	951
Wind	3898	4866	6050	7076
Solar PV	171	1878	3646	6065
Bio	374	375	375	408
Nuclear	1608	1608	1608	1608
Cogen	1322	1709	1710	2106
Total Clean Energy	**20884**	**23947**	**26900**	**30828**
Combine Cycle	25340	27393	30402	35030
Thermal Conventional	12665	12315	11831	11831
Gas Turbine	2960	2960	2960	3793
Engines	739	880	891	949
TV Coal	5463	5463	5463	5463
Total Thermal	**47167**	**49011**	**51547**	**57066**
TOTAL	**68051**	**72958**	**78447**	**87894**

Source: PRODESEN 2020-2034

Table 1.
MX installed capacity by type [MW].

It is expected that in the coming years the aforementioned trend in the expansion of the generation fleet in all LATAM's countries will continue, as a result of which is to be expected growing participation of WIND/SOLAR PV generation in the generation mix that supplies the demand.

3. Reliability in the operation of the electrical system

The LATAM's countries supply the demand with a mix of conventional generation plus a growing share of NCRE generation.

From the point of view of reliability in the supply of demand, countries with a high share of hydro generation (BRA, COL) use the energy stored in reservoirs to provide reserves for rapid management, contributing to the system reliability even against the high intermittence that characterizes the NCRE generation.

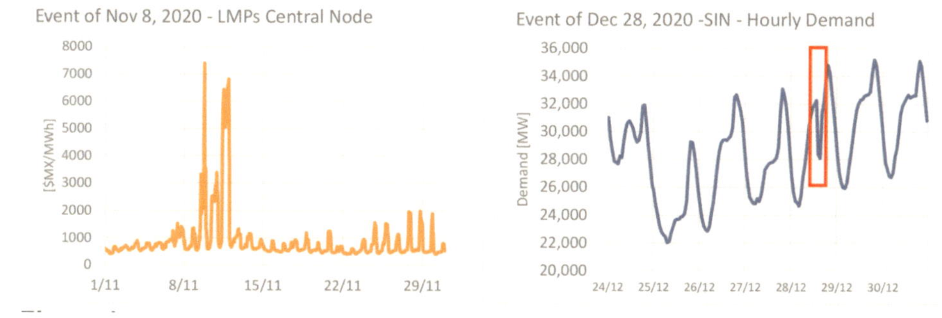

Figure 4.
Events with high volatility in the operation of the market.

On the other hand, countries with a low share of hydro generation must have another type of quickly managed reserves to guarantee the security of supply in the face of sudden changes in the production of NCRE generation.

Table 2 shows the participation of hydro, thermal, and renewables in the generation mix of LATAM's countries.

The MX electrical system has a minimum participation of hydro generation in the generation mix (9% in 2020). In this system, the reliability problems increase because thermal generation is mainly composed of conventional thermal generation (steam turbines) and the available gas turbines are very limited (3700 MW in 2020), so the dispatched thermal generation cannot follow the fast variations of ERNC production.

The aforementioned characteristics of the MX electrical system, together with the growth observed in NCRE generation, were responsible for serious reliability problems registered in two events during 2020.

1. In the week of November 8, high energy prices (LMPs) were recorded (>6000 MXN/MWh, 300 USD/MWh). They are unprecedented values in operation and were not repeated in the remaining days of the year.

2. In the afternoon of December 28, there was strong instability in the power system operation, which resulted in a severe blackout affecting over 10 million people, with a load cutoff of 8696 MW (Source: CENACE6).

Figure 4 show both events. On the left, it is observed very high marginal prices observed in the event of November 8. On the right, it is observed a sudden reduction in the demand in the event of December 28.

1. **Event of November 8.** High LMPs were recorded in the week of November 8, 2020, which are explained by a very low generation reserve, which resulted in the need to dispatch generation units with a very high VPC. It is observed very high marginal cost (LMPs) in hours with minimum wind and solar production.

2. **Event of December 28.** On the afternoon of December 28, there was strong instability in the operation of the system, which resulted in a significant load cutoff (8696 MW). The instability recorded in the power system was due to low

Country	Hydro	Thermal	Renewable	Total (GWh)
Argentina	22%	69%	9%	134,177
Bolivia	32%	63%	5%	9212
Brazil	72%	17%	11%	557,055
Chile	27%	54%	20%	77,567
Colombia	72%	27%	1%	69,323
Costa Rica	72%	0%	28%	11,534
Ecuador	89%	9%	2%	27,301
El Salvador	37%	17%	47%	5386
Guatemala	52%	21%	27%	11,122
Honduras	25%	42%	32%	9578
Mexico	**9%**	78%	13%	247,415
Nicaragua	15%	29%	55%	3748
Panama	48%	40%	12%	5420
Perú	60%	35%	5%	49,187
Uruguay	34%	7%	59%	11,596

Table 2.
LATAM's countries—generation by technology (2020).

availability of thermal generation, as a result of which, upon a fault in the transmission system, the power system had not enough generation capacity (reserve), which led to a frequency drop in the system with the subsequent load loss to restore the dynamic balance of the power system. **Figure 5** shows the production of renewables during the event of instability mentioned. The figure includes the marginal cost (local marginal prices—LMPs). During the week, it is observed high variations in renewable production (5000 MW of solar production and 3000 MW of wind generation) resulting in very high variations of marginal prices that show low reserve in the system.

As a result of the aforementioned events, the Government of President Andrés Manuel Pérez Obrador (AMLO) enacted new rules for market operation meant to preserve the power system's reliability, safety, and continuity. The most relevant are as follows:

1. Include limits for the integration of NCRE into the electric system.

2. Restrictions on the dispatch of the NCRE generation (curtailment).

3. To provide Ancillary Services, the ISO is enabled to include in the generation dispatch some proportion of mandatory generation (must-run).

4. NCRE generators must be able to control the voltage at their connection node at all times.

5. NCRE generators do not provide firm energy to the electrical system.

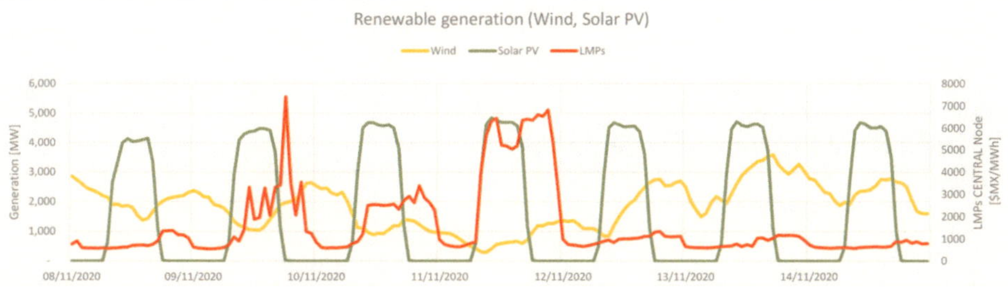

Figure 5.
Hourly NCRE generation (wind, solar PV). Source: Own production based on CENACE, MX.

The new rules introduce severe restrictions to the integration of NCRE generation in the system, resulting in higher energy prices and higher greenhouse gas production as a result of increased production of thermal generation replacing NCRE generation.

4. Market rules to incentive NCRE integration together with a safe system operation

4.1 Generation dispatch programming

The determination of the hourly production of each generator that participates in the market results from an operation programming that covers different time intervals from 1 day, 1 week, and the medium/long term. This makes it possible to guarantee that the programming of the operation meets the objective of supplying the demand with adequate quality service at a minimum cost within the time horizon covered by each program [7].

In electrical systems with high participation of hydro generation in the generation mix, the energy that is stored in the reservoirs is used to supply the demand, thus avoiding fuel costs in the thermoelectric units. However, the availability of hydro energy is limited by the storage capacity in the reservoirs, which also affects the operational safety of the system.

This introduces a dependency between today's operating decision and future operating costs, including possible costs due to insufficient generation capacity resulting in demand cuts (non-supply energy).

If the existing hydro reserves are used in the short term to minimize thermal costs and a severe drought occurs in the future, high-cost rationing could occur, affecting the quality of service.

If, on the other hand, the hydro reserves are not used in the short term, through more intense use of thermal generation, and the future hydro inflows to reservoirs are high, water spill may occur, which represents a waste of energy and, consequently, an increase in the operating cost of the electrical system.

The optimal use of the water stored in the reservoirs corresponds to the point that minimizes the sum of the total costs incurred (present plus future costs). As shown in **Figure 6**, the hydro production that minimizes total cost is where the derivatives of the present cost (FCI) and future cost function (FCF), with respect to the volume of water stored in the reservoirs, are equal with opposite signs.

$$\frac{\partial (\textit{Total}\ \text{Cost})}{\partial V} = 0 \tag{2}$$

where
V [m^3]: Stored volume of water in the reservoir.
Then

$$\frac{\partial}{\partial \text{V}}\text{FCI} = -\frac{\partial}{\partial \text{V}}\text{FCF} = \text{Water Value} \tag{3}$$

The last equation says that minimum total cost is achieved when the reservoir reaches a level where the marginal immediate cost of using water is equal to the marginal future cost of using water now (with a different sign).

Any of both derivatives are known as WATER VALUE [$/$Hm^3$] and represent the "opportunity variable cost" of the stored water.

Immediate cost function (FCI) is directly calculated as the least-cost thermal complement to hydro energy production. The future cost function (FCF) is conceptually calculated through the simulation of future system operation and the calculation of corresponding operating costs.

Due to the variability of inflows to reservoirs, which fluctuate seasonally, regionally, and from year to year, this simulation is carried out on a probabilistic basis, that is, using a large number of hydrological scenarios (historical data; dry, medium, and wet years).

SIMULATION = > FUTURE COST FUNCTION.

FUTURE COST FUNCTION = > HYDRO OPERATION POLICY FOR SHORT TERM.

The optimization of generation resources in a hydrothermal system, such as the ones existing in LATAM's countries, requires the use of mathematical optimization models that simulate the hierarchical decision-making process that must be carried out (strategy, tactics, and operation).

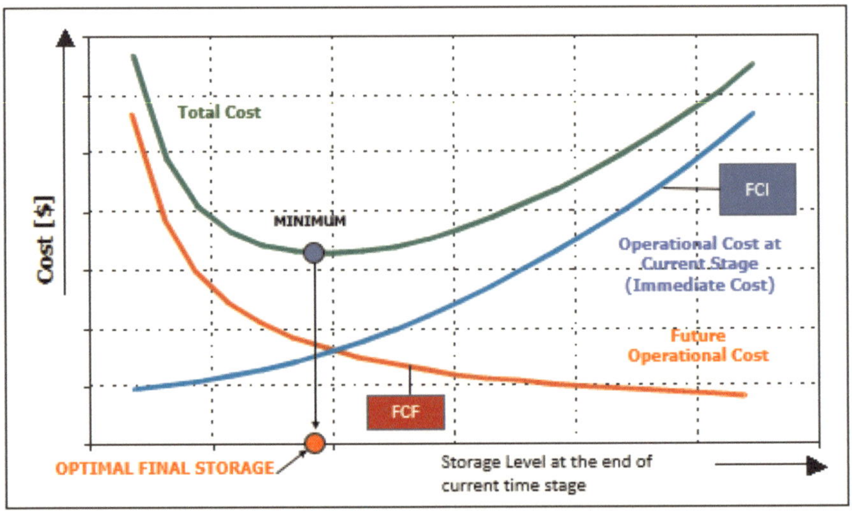

Figure 6.
Hydro production optimization.

The generation dispatch is the result of sequential processes of calculation that divide the total problem, in time and space (multiples reservoirs), based on solvable models in reasonable computational times, which are coupled to each other through the conditions of the border that joins them [8].

The level of detail of the modeling, in time and space, is a compromise between the information required, the complexity of the mathematical problem, and the computational resources available.

Thus, planning the operation of hydrothermal systems can be divided into two major steps, which are as follows:

- **Planning:** Related to the simulation of the "optimal operation" in the coming months/years. For planning, it is used an **LTP** model.

- **Programming:** Related to the programming of the "optimal operation" in the short term (next few hours, typically on a week). For programming, it is used an **STP** model.

The optimization models related to both processes are quite well-known and are generally used in one way or another by electricity system operators around the world [9–11].

Given the differences between the LTP and STP models, each one of them is executed and solved independently (at different times) (**Figure** 7).

The short-term optimization is carried out using the boundary conditions established by the LTP model for the $\mathbf{T_{ST}}$ time, which determines the end of the period for which you want to schedule operations with the STP model, called short-term scheduling that covers period {1, $\mathbf{T_{ST}}$}; the LTP model covers the period {$\mathbf{T_{ST}}$ + 1, **T**} that needs to be analyzed to determine the future implications of decisions made in the present, {1, $\mathbf{T_{ST}}$}.

Frontier conditions between both models can be established at least in one of the following ways:

- **Through economic variables (dual/prices):** Based on the setting, the opportunity price of the stocks at the end of period {1, TCP}, in this case, the hydro resource (water stored); it also applies to other storable resources such as the unused amounts of previously purchased fuel (take or pay contracts).

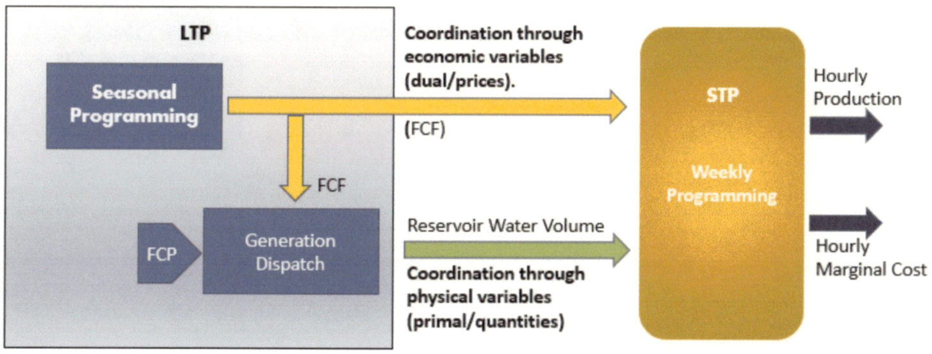

Figure 7.
STS and LTS models coordination.

- **Through physical variables (primal/quantities):** Based on the setting, the amounts of resources available to be used during the operating period {1, TCP}.

Then, the minimum simulation period for the long-term programming is different for each system, according to their total water regulation capacity:
Brazil => 4 years
Chile => 4 years
Colombia => 2 years
Argentina => 1 year
Ecuador => Seasonal

4.2 Energy marginal cost

The generation economic dispatch results in the production of each power plant in each time interval (1 hour).

In each hour, the energy marginal cost is equal to the VPC of the generator with the highest VPC that, in each hour, is producing energy according to the results of the economic dispatch. The energy prices in the spot market are equal to the energy marginal cost [12].

Figure 8 show historical energy prices for MX Market (Local Marginal Prices (LMPs) in the Day-Ahead Market (DAM)) at the Hermosillo node.

The marginal costs of energy have intraday and seasonal variations. The intraday variations, with a time step of 1 hour, are mainly due to (i) the hourly variations of the system demand (typically, the system demand is maximum in the evening/night hours and minimum demand in the early morning hours), (ii) the production of solar generation that reduces the energy prices at solar hours, and (iii) any time, by the effect in the power balance resulting from the forced outage of a generation unit or a transmission line.

Within the year, the average marginal costs of energy vary mainly due to the added effect of (i) seasonal variations in demand (in MX typically the demand in the summer months is greater than that of the winter months), (ii) due to the effect of

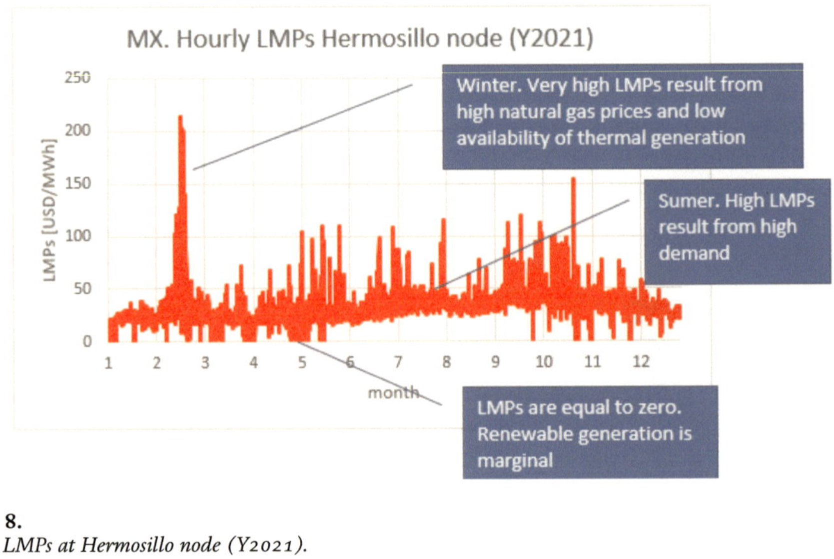

Figure 8.
Hourly LMPs at Hermosillo node (Y2021).

rainfall that modifies the water inflows that reaches the reservoirs of the hydro plants, and (iii) seasonal changes in fuel prices.

If at some moment there is not enough generation availability to supply the demand, including the reserve margin for frequency regulation, the energy marginal cost result equal to the cost of not-supply energy (CNSE). This results in a strong economic signal toward improving the availability of generating units and thereby improving the security of the system.

The growing addition of new intermittent generation capacity (wind and solar) will produce variations in the generated power that can be significant in time intervals of less than 1 hour. In these operating conditions, for the marginal cost of energy to produce efficient economic signals to promote the availability of generation, it will possibly be necessary to determine the marginal costs of generation in time intervals of less than 1 hour [13, 14], mainly in markets with low hydro participation (like Mexico).

As a result, greater volatility is expected at energy prices in the spot market due to variations in the marginal costs of energy resulting from the intermittency in the production of NCRE generators.

To mitigate the risks of high prices in the spot market, supply contracts, with freely agreed conditions between generators and consumers, are efficient commercial instruments that allow:

- Stabilize prices (price hedging)
- Facilitate investment in generation
- As a mechanism for long-term supply guarantee
- Promote adequate generation availability

Note: In LATAM's electricity markets, the supply contacts are **financial contracts** (not physical contracts). This means that the generator supplies, each hour, the contracted energy with its production, if it is dispatched, or by buying the contracted energy in the spot market if it is not dispatched, so financial contracts do not modify the economic generation dispatch.

4.3 Day-ahead markets and intraday markets

At the international level, there are several examples of wholesale energy markets with two or more settlement instances. An example is the wholesale electricity market of Mexico, where a day-ahead market (DAM) and a real-time market (RTM) operate. The other LATAM's markets are all markets that operate only in real-time.

In markets that operate only in real-time, the income of the generators in each hour ($GEN(h)) results from valuating the energy generated (EG(h)) at the marginal cost of the hour (CMgR(h)).

$$\$GEN(h) = EGR(h) \times CMgR(h) \tag{4}$$

In the markets that operate with two or more settlement instances, for example, MX where there is the DAM and the RTM, the income of the generators is determined by the following expression:

$$\$GEN(h) = \underbrace{EGA(h) \times CMgA(h)}_{\{1\}} + \underbrace{[EGR(h) - EGA(h)] \times CMgR(h)}_{\{2\}} \tag{5}$$

In this case, the income of the generators has two terms, which are as follows:

- The term {1} corresponds to the income resulting from selling the energy expected to be generated (EGA) at the marginal cost of the DAM (CMgA).

- The term {2} corresponds to valuing the error of the forecast generated energy (EGR-EGA) at the marginal cost in the RTM (CMgR).

Therefore, in markets where there are at least two settlement instances (DAM and RTM), the generators are exposed to the uncertainty of the marginal costs in the RTM due to the difference between the expected energy to be generated (offer to the DAM) and the energy real generated in the RTM.

This exposure generates risks that generators seek to mitigate by improving their production forecasting tools for the following day. This helps to mitigate the adverse effects of the volatility of the production of NCRE generators on the security of supply. In markets with multiple intraday settlements (like European electricity markets), this is improved due to the probable production in real-time being better known.

In summarizing, the advantages of multiple settlements are: (i) allows generators to hedge the risks associated with real-time price volatility; (ii) reduces operating costs by allowing the market operation to be better scheduled to meet the demands at a minimum cost; and (iii) allows programming the operation of resources to deal with the uncertainty of large-scale variable renewable generation.

Given the existing structural differences between the different electricity markets (e.g., due to the generation mix that supplies the demand), the implementation of the multiple settlements scheme requires special attention to analyze the effect that these have on the generators' incentives to have high-precision forecasts, the opportunities for better coverage of risks due to price variability in the real-time market, and the incentives to install flexible generation resources that are capable of adjusting their dispatches at short notice. Other aspects to be evaluated are the number of settlements that it considers in the short-term market, the products that are traded in the multiple settlements (energy, ancillary services), and the sophistication of its matching models.

By way of background, it should be noted that markets in the United States are typically characterized by a day-ahead market and a real-time market (i.e., two-settlement markets), with co-optimization of energy offers and frequency reserves, and with matching algorithms from the previous day's market that consider unit commissioning restrictions and their associated costs, in addition to transmission system restrictions, based on a so-called Security Constrained Unit Commitment (SCUC). On the other hand, wholesale markets in Europe are characterized by the sequential acquisition of reserves and energy, where energy is traded in auctions the day before and multiple instances of intraday auction markets or continuous bilateral transactions [15]. In this case, the daily and intraday market instances do not consider security restrictions and transmission corridor congestion, which are evaluated and managed by each system operator, generating deviations in the real operation for market matching.

4.4 Forecast of the NCRE generation production profile

To schedule the market operation, the market operator needs to forecast the probable production of NCRE generators. NCRE generation forecasting is a rapidly evolving field [16, 17].

With the expected increase of NCRE generation, including generation within the distribution networks, the problem of forecasting the production is complicated since usually there is no information on the NCRE produced within the distribution networks. In these cases, the net demand of the distribution system must be forecast, which is equal to the consumed demand minus the existing NCRE generation in each distribution system.

To increase reliability in the operation of the market, regulatory changes should therefore be introduced to improve the quality of information on the expected production profile of NCRE generators. Improving the forecasts should be the combined responsibility of the system operator and the market agents. Departures between forecast and actual values should give rise to economic incentives to improve forecasts, for example, via intraday markets above described.

4.5 System reserves

The generation resources that provide flexibility to the electrical system are typically the following:

- Conventional generators (quick start thermal, hydro),
- Spin reserves (primary, secondary)/ramp,
- Load reduction,
- Energy storage systems,
- FACTS equipment as a means of controlling transmission capacity that improves the ability to share reserves between areas of the electrical system,
- Interconnections with neighboring systems (other countries).

The development of the aforementioned technologies/schemes in the electricity market may result from the economic signals produced by the market. When this is not possible (lack of competition, oligopoly), regulated mechanisms (Ancillary Services) are created to make the development of the necessary technologies economically viable to achieve a safe and minimum-cost operation of the electrical system.

The amount of reserve (i.e., for frequency regulation) required in the system results from a trade-off between the cost of the reserve and the resulting quality of service.

The determination of the required reserve is carried out through reliability studies [18–20] that result in reliability indices (e.g., energy not supply) associated with insufficient generation for different values of power reserve. Based on the results, the cost of the reserve and the cost of the quality of service are determined. The optimal reserve is the one that allows minimizing the sum of the costs of the reserve plus the costs of the NSE due to insufficient generation.

4.6 Optimal reserve power determination

The optimal reserve power required by the electrical system is determined through reliability studies that allow knowing the quality of the supply of the demand based on the reserve power in the electrical system.

For example, the quality index could be the ratio between the NSE and the total energy supplied, or the probability of loss of load, both due to the effect of forced contingencies in the generation fleet. In what follows, these quality indices are identified as RE (**RE**liability Index).

Electrical systems are composed of a large number of generating units, each with its own technical and availability characteristics, including the availability of primary resources in the case of renewable power plants (inflow water, wind speed, and solar radiation levels), and where demand varies from time to time following generally known and repetitive patterns. In these kinds of systems, and considering a very low allowed NSE (in the range of 1×10^{-3} to 1×10^{-5} of the served energy), the system quality index RE is typically an exponential function of reserve power (PRES) required to control the quality of the supply. The NSE is proportional to RE index.

$$RE \equiv e^{-k \times PRES} \tag{6}$$

$$NSE = k_1 e^{-k \times PRES} \tag{7}$$

where

k, k1: Constant that results from system reliability studies.

PRES [MW]. System reserve power.

The total cost incurred in the electrical system results from the sum of the cost of the energy not served that results in a certain quality index RE, plus the cost of providing the reserve power that allows obtaining said quality level.

$$Total\ Cost\ [\$] = CNSE \left[\frac{\$}{MWh}\right] \times NSE[MWh] + CRES \left[\frac{\$}{MW}\right] \times PRES[MW] \tag{8}$$

where

CNSE: Unitary cost of the NSE.

NSE: Not supply energy resulting from reliability analysis.

CRES: Unitary cost of the reserve power.

PRES: Reserve power.

In the optimum

$$\frac{d\ (Total\ Cost)}{dPRES} = 0.0 \tag{9}$$

That allows to obtain the optimum value of System Reserve Power (PRESopt):

$$\frac{d\ (NSE)}{dPRES} = \alpha = -\frac{CRES}{CNSE} \tag{10}$$

$$PRES_{opt} = \frac{\ln\left(kk_1 CNSE/CRES\right)}{k} \tag{11}$$

Figure 9 shows the described optimization process. The red curve shows the variation of the non-supply energy (NSE) as a function of the reserve power (PRES). The higher the reserve, the lower the NSE. The optimal PRES is the one that meets the condition that the derivative of the NSE, with respect to PRES, is equal to alpha (α).

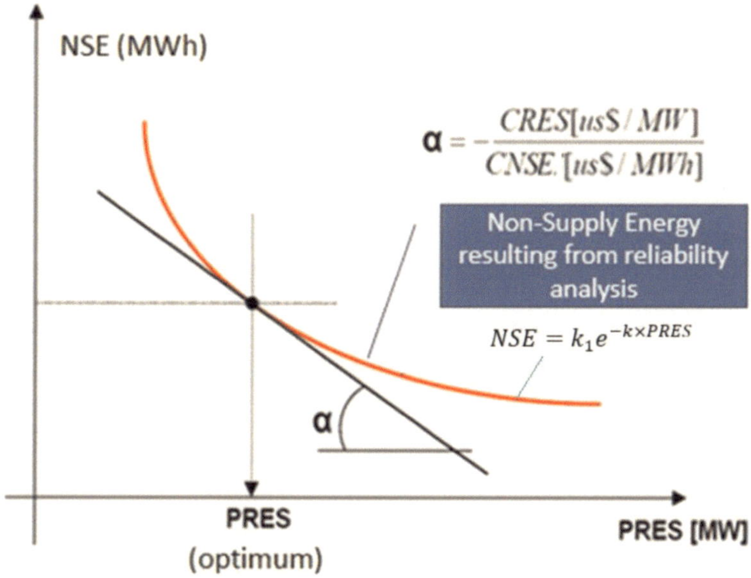

Figure 9.
Optimal power reserve.

4.7 The cost of non-supply energy (CNSE)

As demonstrated in the previous point, the reserve power (PRES) required by the electrical system depends on the cost of not supply energy (CNSE), with the reserve power being greater the higher the CNSE.

The CNSE concept includes a group of economic costs that can affect society, as a whole, when the supply of electricity cannot be provided to the extent required by consumers. The NSE is the amount of energy potentially demanded (presumed energy) that cannot be supplied.

In the commodity markets, in the absence of enough supply, the price of the product increases and the quantity demanded adjusts automatically (elasticity), first withdrawing those consumers with lower utility or consumer surplus, which is economically efficient, thus minimizing the reduction in societal benefit.

However, the electricity sector has certain special characteristics, due to which the CNSE concept is used, instead of the more natural idea of balance between supply and demand:

- Electrical energy, as cannot be stored, if there is not enough supply available in the electrical system, and the excess demand is not interrupted, the system runs the risk of collapsing.

- Electricity demand is very inelastic, which is why price signals maybe not be enough to return the system to a situation of supply/demand balance.

The valuation that consumers make of the NSE, in general, needs to be estimated. The economic costs that can affect society as a whole when electricity supply is not enough are of various kinds. The main difficulties that arise in estimating the CNSE are:

- The complexity of modeling the link between the not enough supply and the impact on people's well-being, as well as on the economic activities that are affected, depends on the type of interruption.
- The CNSE varies depending on the existence or not of selectivity in the cuts, or whether they affect all consumers equally or not.
- The complexity of establishing the acceptable limits of deterioration in the quality of supply compared to the alternative of interrupting demand.

There is a wide variety of methods that can be used to calculate the CNSE. There are two large families of approaches, which are as follows:

- **Indirect methods:** Aggregated data are used, for example, the relationship between energy demand and GDP, or based on the cost of self-generation, or through the Leisure Work Exchange Theory [21].
- **Direct methods:** They are based on conducting surveys or interviews with consumers to inquire directly about the economic and noneconomic effects linked to the interruption of electricity service.

The CNSE is significantly reduced if the interruption condition can be anticipated by the consumer in such a way that they can take precautions in the event of a probable supply interruption. To take this aspect into account, the CNSE is determined for short-term failures (when the failure condition is not prevented) and long-term failures (when the failure is scheduled and known by the consumers).

4.8 Capacity balance market

The purpose of the **Capacity Market** is to provide an economic signal to incentivize the installation of enough generation capacity to supply demand while satisfying defined reliability criteria [22].

If the generation fleet was designed and operated to supply the system demand at a minimum total cost (the sum of investment costs (CAPEX) plus OPEX operating costs), it is fulfilled that all generators cover all of their investment costs and operation (sufficiency principle) if all generators receive as remuneration: (i) a payment for energy ($ENE), which results from valuing their production at the Marginal Cost of the Energy in the market, and (ii) payment for firm capacity ($POT), which results from valuing their firm capacity at the price of capacity (PPOT) in the power balance market.

The value of PPOT is determined in most of LATAM's markets as the annual fixed costs (CAPEX, fixed O&M costs) of a Turbo Gas-type power plant with an installed capacity approximately equal to the annual growth of the maximum demand of the system.

The payment for capacity that generators receive has a direct effect on the reliability of the electrical system.

With capacity payments, generators require lower marginal rent (spark spread) to cover their capital costs (CAPEX). So, for the same demand, lower marginal rent results from lower market marginal costs that result from a higher generation availability and consequently lower non-supply energy (NSE) probability.

The firm capacity of the generating units depends on their technology. For hydroelectric plants, firm capacity is determined typically as the power generated in the hours of maximum demand (or minimum reserve) for a very dry hydrological condition.

For thermal power plants, firm capacity is usually equal to their average available capacity. Large thermal power plants, compared to the supplied demand, produce large disturbances in the operation of the electricity market when they go out of operation due to an unscheduled event (failure). When the power of the failed plant exceeds the reserve margin for frequency regulation, a load cut will be necessary to balance supply/demand, a situation that could lead to a massive load cut. In such a situation, the firm capacity of large thermal plants could be reduced as an economic signal that shows the impact on the quality of service of the system.

As an example, **Table 3** shows the installed capacity and forced output rate (FOR) of a fictitious generation fleet made up of 10 generating units (G#1 to G#10). One of the generators (G#3) has an installed capacity (300 MW) equivalent to 25% of the total installed capacity of the generation park.

Figure 10 shows the total power available for a given probability of exceedance resulting from considering all the possible operating states (2^{10}) of the generation park considering the operating states of each generation unit.

Generator	G#1	G#2	G#3	G#4	G#5	G#6	G#7	G#8	G#9	G#10
Installed Capacity [MW]	100	150	300	50	20	80	120	200	20	180
Forced Output Rate	20%	15%	20%	30%	50%	35%	15%	10%	20%	25%
Available Capacity [MW]	80	127.5	**240**	35	10	52	102	180	16	135

Table 3.
Example of the generation fleet.

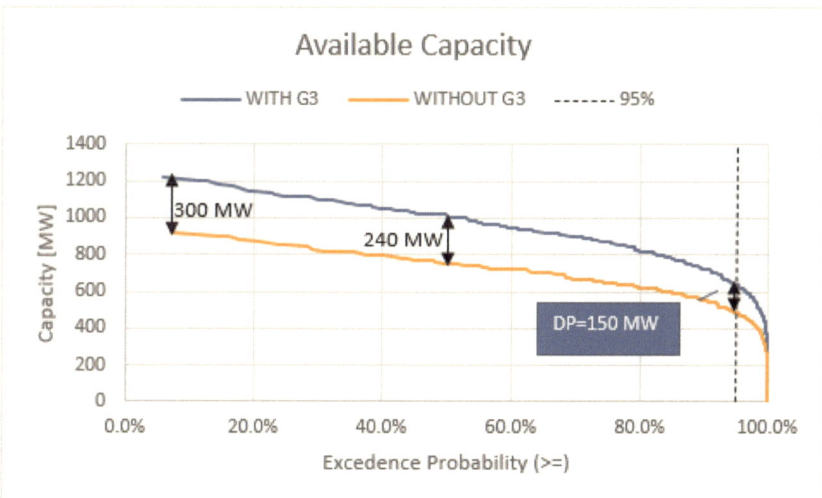

Figure 10.
The firm capacity of thermal plant.

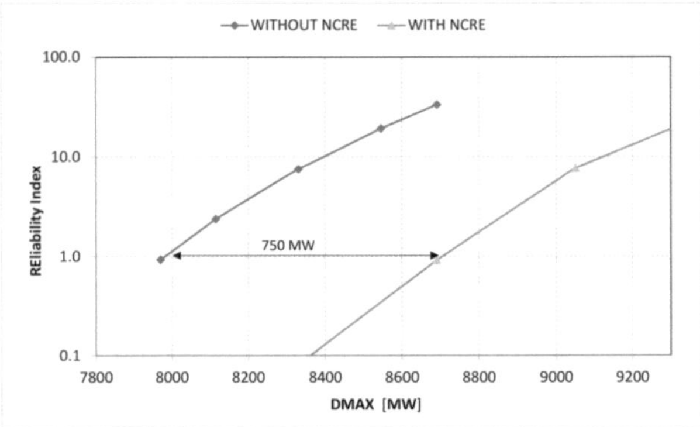

Figure 11.
Effective load carrying capability of NCRE.

Two curves are presented, the blue curve considering the entire generation park and the red curve WITHOUT generator G#3.

It is observed that for a probability of exceedance of 95%, generator G#3 contributes only 150 MW, while its average available power is 240 MW, so the firm capacity of generator G#3 should be reduced to 150 MW.

Up to now, in the LATAM's markets, there is no consensus on the firm capacity of NCRE plants. The main reason is that it cannot be guaranteed that this type of generator will produce energy during the hours of minimum reserve of the system, which are typically at night hours.

In markets with a significant share of hydro generation in the generation mix that supplies demand, such as most of the LATAM's markets, it is observed through reliability studies that with NCRE generation, a greater demand can be supplied without compromising the quality of service. This is justified because the production of NCRE generation allows a greater volume of water to be stored in the reservoirs, which allows a greater hydro generation in hours of minimum reserve.

As an example, **Figure 11** shows the evolution of a reliability index for different levels of demand in the Peruvian market. Two curves are presented, WITH and WITHOUT NCRE generation (installed capacity 1100 MW). It is observed that WITH NCRE generation, demand can be increased by 750 MW without reducing the reliability index, which shows that in the Peruvian system, NCRE generation effectively contributes to improving the quality of service in supplying the demand [23–25].

In electrical systems with high participation of hydro energy, the hours where there is a greater risk of not being able to supply the demand is at evening/night hours when the demand is usually maximum. In these systems, the firm capacity result from the generation available in this hour range.

On the other hand, in systems with low participation of hydro generation, the supply risk is at any time of the day and even more so with significant participation of NCRE generation in the generation mix, as occurs in the MX electrical system [26].

For this reason, in the MX electrical system, the firm capacity of an NCRE generator results from the generator production in the so-called critical hours, the 100 hours of minimum reserve of the year. **Figure 12** shows the system generation reserve, in MX, in each hour of 2019, in red the 100 critical hours. The firm capacity of the NCRE generators is measured in these critical hours.

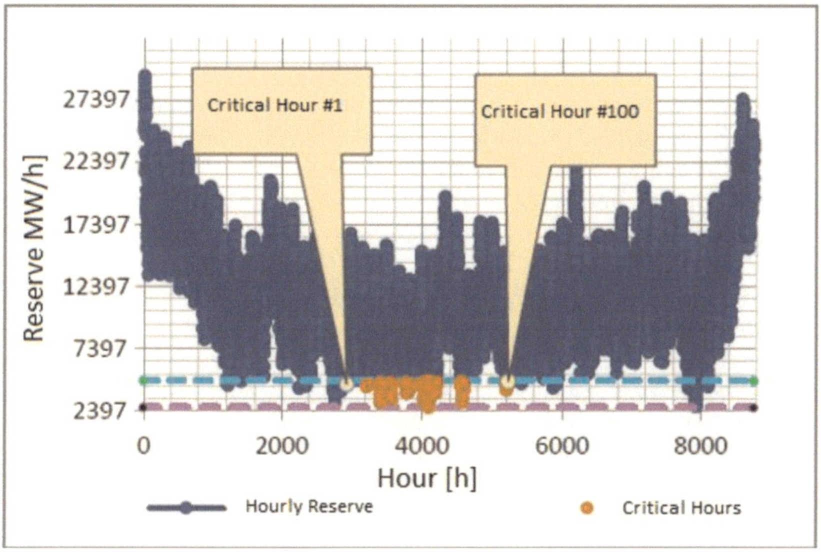

Figure 12.
MX capacity balancing market – 100 critical hours, 2019 year.

4.9 Storage energy systems

Storage energy systems are a set of technologies and operating methods that allow energy to be conserved for later use (similar to what happens with hydro plants). Energy storage is currently based on a broad set of technologies, many of which already have a solid state of maturity, while others are less consolidated, which require progress in some aspects and improve their performance, costs, and competitiveness.

Examples of existing storage systems in LATAM's countries are (i) batteries (BESS), in which the charging phase is carried out by storing chemical energy in the batteries, and (ii) pumping hydro plants in which during the charging phase it accumulates water in the reservoir (potential energy) by pumping, which is transformed into electrical energy during the discharge phase. The charging phase is usually fed with energy withdrawn from the same network into which the accumulated energy will be injected.

Of the storage technologies, it is worth highlighting the high potential for the development of BESS, which constitutes an effective complement to NCRE generation for the safe supply of demand at a minimum cost. Other possible uses of BESS concerning the safety and quality of the energy supply are as follows:

1. Frequency control reserves to manage contingencies, especially with a relatively large proportion of NCRE in the system;

2. Voltage stability in the power system;

3. Provide peaking capacity, deferment of new generation capacity, and displacement of thermal generation;

4. Enhanced system security by supplying energy during shortages in electricity generation.

The NCRE generation technically today can offer operating reserves for frequency regulation. However, it must be taken into account that since NCRE generation has zero production opportunity cost (the energy that is not produced is lost) by requiring a power reserve margin to offer the regulating band, the provision of reserves normally has a high operating cost results from the difference between the price of spot energy and the variable cost of the generator, which is zero. This margin is usually much lower in other technologies, so from the economic point of view, it is better to assign the frequency regulation reserve to conventional power plants or storage systems.

Storage generators, particularly batteries, can be a lower-cost alternative for NCRE generators to provide power reserves for frequency regulation.

NCRE generators can also increase their firm capacity by adding batteries to their installed generation capacity, charging the batteries with their generation at hours when there is a high reserve margin (low marginal energy prices), and injecting the stored energy into the electrical system at hours of the day when the reserve is minimal (high marginal prices). This configuration is called a "hybrid generator."

The storage generators can also operate in the electricity markets independently of other generators in the system, under the so-called "stand-alone" configuration. In this case, the owner of the storage media (battery) covers the investment costs via the marginal rent that results from the storage process in hours of low marginal cost and the sale of the energy stored in hours of high marginal cost. The difference between the marginal costs during loading and unloading of the storage medium must cover at least the losses in the process of loading/unloading the storage medium.

A stand-alone storage generator can also be remunerated by its firm capacity. In this case, system operation studies must demonstrate that it will be economically convenient to charge the storage system when the electrical system is in a critical situation with a minimum reserve margin. The economic feasibility will depend on the difference between the marginal costs between the loading and unloading processes when the electrical system has a minimum reserve. If the difference in marginal cost does not cover the cost of losses in the storage system, so the firm capacity will be equal to zero.

4.10 Demand participation

The demand can participate directly in the spot market via offers to withdraw demand if the price of energy in the spot market exceeds the offered price. To this effect, loads should have the technical capacity to respond to the ISO instructions to reduce the load in real time. From the point of view of the economical minimum cost dispatch, the demand withdrawal offers are similar to a thermal plant with negative capacity and a VPC equal to the offered price.

Consumers can also play an active role in the system's security of supply via their response to the electricity rates paid to the distribution company from which they buy energy. In many cases, the rates include high charges if the time in which the consumer has his maximum demand coincides with the time of day in which the aggregate demand of the distribution company is maximum. In these cases, the consumers, mainly industries, reduce their consumption, so, their demand is elastic to electricity rates. This can be seen in **Figure 13**, which shows the hourly demand of a typical day for large consumers in the Peruvian market. In the hours where rates are high, 6–10 pm, demand is significantly reduced. This lower demand, at times when the system

DOI: http://dx.doi.org/10.5772/intechopen.107062

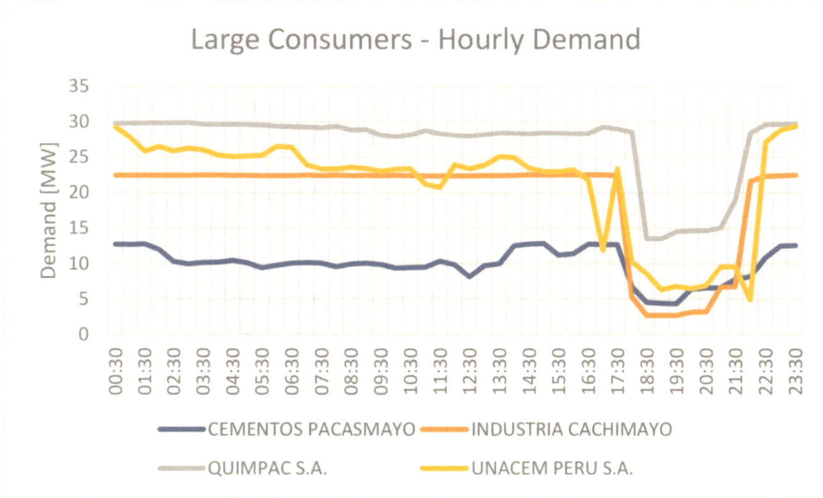

Figure 13.
Peruvian electricity market. Hourly demand of large consumers.

typically has minimal reserves, contributes to increasing system reserves and thus improving the security of supply.

5. Conclusions

In the 1990s, wholesale electricity markets were organized in the LATAM's countries to introduce competition in the supply of electricity as a way of supplying demand with an adequate quality of service at a minimum cost.

The initial design of these markets mainly considered that the demand would be supplied with a mix of hydro and thermal generation.

In recent years, the growing participation of NCRE energies, mainly wind and solar-PV, in the generation mix has been observed, promoted by a significant reduction in the capital costs of these technologies and the need for countries to advance in the replacement of thermal generation, mainly the generator based on coal, to mitigate the effects of climate change and thus move toward a production matrix-based mainly based on clean energy (energy transition).

The greater participation of solar and wind energy, characterized by variable generation depending on climatic conditions (wind level and solar radiation), introduced a strong instability in the electrical systems that put at risk the normal supply of demand in safe conditions.

To mitigate these effects, the rules of the wholesale electricity markets of the LATAM's countries include, or their implementation is being evaluated soon, strong economic incentives that seek to maximize the availability of energy at times of minimum generation reserve. In this sense, the following stand out:

- Generation dispatch, based on reliability constraints and minimum cost.

- Systems reserves (frequency regulation and voltage control) are dimensioned to supply the demand at minimum cost including CNSE.

- Energy prices are equal to the generation short-term marginal cost including the CNSE when there is not enough energy to supply the demand.
- Anticipated energy markets (day ahead), to promote better information of expected ERNC production for the generation economic dispatch.
- Supply contracts, to mitigate the risks of high prices in the spot market and to promote generation long-term expansion that improves system security.
- Firm capacity market, to promote a reliable operation by remunerating the contribution of the power plants based on their contribution to the security of supply.
- Distribution rates that promote consumer participation reduce its demand in hours of minimum reserves.

The electricity markets of LATAM's countries showed strong dynamism from the beginning, mainly due to private participation in the electricity generation segment.

It is estimated that this behavior will continue in the future, for which it will be necessary for market regulations to evolve to allow the integration of new technologies efficiently, thus contributing to achieving the objectives of the energy transition without compromising the secure supply of demand.

Author details

Daniel Llarens[1,5]*, Laura Souilla[1,2,5], Santiago A. Masiriz[3,5] and Gastón R. Lestard[4,5]

1 La Plata National University, Argentine

2 MA in International and Development Economics, Yale University, United States

3 MBA Torcuato Di Tella University, Argentine

4 Buenos Aires University, Argentine

5 Grupo Mercados Energéticos, GME-Global, Argentine

*Address all correspondence to: dllarens@gme-global.com

References

[1] GE Energy. Western Wind and Solar Integration Study (WWSIS). NREL/SR-550-47434. Golden, CO: National Renewable Energy Laboratory. Available from: www.nrel.gov/docs/fy10osti/47434.pdf; 2010. [Accessed: September 2013]

[2] Puga N. The importance of combined cycle generating plants in integrating large levels of wind power generation. The Electricity Journal. Available from: www.bateswhite.com/media.pnc/4/media.344.pdf. 2010. [Accessed: September 2013]

[3] Western Governors' Association (WGA). Meeting Renewable Energy Targets in the West at Least Cost: The Integration Challenge. Denver, CO: Western Governors' Association. Available from: https://westgov.org/component/joomdoc/doc_details/1610-meeting-renewable-energy-targets-in-the-west-at-least-cost-the-integration-challenge-full-report; 2012. [Accessed: September 2013]

[4] IRENA. Planning for the Renewable Future. International Renewable Energy Agency. 2017. ISBN 978-92-95111-06-6

[5] Rudnick H, Barroso LA, Llarens D, Watts D, Ferreira R. Flexible connections: Solutions and challenges for the integration of renewables in South America. IEEE Power and Energy Magazine. 2012;**10**(2):24-36. DOI: 10.1109/MPE.2011.2178300

[6] Lazard. Levelized Cost of Energy, Levelized Cost of Storage, and Levelized Cost of Hydrogen. 2021

[7] Gonzalez-Romero IC, Wogrin S, Gomez T. Transmission and storage expansion planning under imperfect market competition: Social planner versus merchant investor. Energy Economics. 2021;**103**:105591

[8] Available from: https://www.psr-inc.com/softwares-en/?current=p4028

[9] Pereira MVF, Pinto LMVG. Stochastic optimization of a multireservoir hydroelectric system: A decomposition approach. 1985. DOI: 10.1029/WR021i006p00779

[10] Barroso LA, Granville S, Oliveira GC, Thomé LM, Campodónico N, Latorre ML, et al. Stochastic Optimization of Transmission Constrained and Large Scale Hydrothermal Systems in a Competitive Framework. Vol. 2. Toronto, ON, Canada: Proceedings of the IEEE General Meeting 2003; 2003

[11] Lew D, Milligan M, Jordan G, Piwko D. The value of wind power forecasting. In: Prepared for the American Meteorological Society Annual Meeting. NREL/CP5500-50814. Golden, CO: National Renewable Energy Laboratory. 13 pp. Available from: www.nrel.gov/docs/fy11osti/50814.pdf; 2011. [Accessed: September 2013]

[12] Schweppe FC, Caramanis MC, Tabors RD, Bohn RES. Pricing of Electricity. New York, NY, USA: Springer; 1988. Available from: https://link.springer.com/book/10.1007/978-1-4613-1683-1

[13] Milligan M, Kirby B, King J, Beuning S. The impact of alternative dispatch intervals on operating reserve requirements for variable generation. In: 10th International Workshop on Large-Scale Integration of Wind and Solar Power Into Power Systems as Well as on Transmission Networks for Offshore Wind Power Plants Proceedings. Langen,

Germany: Energynautics GmbH; 2011. pp. 197-202. OSTI: 21594641 TRN: DE12GA158

[14] Milligan M, Clark K, King J, Kirby B, Guo T, Liu G. Examination of Potential Benefits of an Energy Imbalance Market in the Western Interconnection. Aarhus, Denmark: NREL Technical Report; 2013. Available from: www.nrel.gov/docs/fy13osti/57115.pdf

[15] Kath C. Modeling Intraday Markets under the new Advances of the Cross-border Intraday Project (XBID): Evidence from the German Intraday Market. 45141 Essen, Germany: University of Duisburg-Essen. DOI: 10.3390/en12224339

[16] Jones LE. Renewable energy integration. In: Practical Management of Variability, Uncertainty, and Flexibility in Power Grids. Elsevier Inc.; 2014. ISBN: 978-0-12-407910-6, ISBN: 9780124081222

[17] Lew D, Brinkman G, Ibanez E, Hodge B-M, Hummon M, Florita A, et al. Western Wind and Solar Integration Study Phase 2 (WWSIS-2). NREL/TP-5500-55588. Golden, CO: National Renewable Energy Laboratory; 2013

[18] Al-Shaalan AM. Reliable Evaluation of Power Systems. London, UK: IntechOpen; 2019. DOI: 10.5772/intechopen.85571

[19] Werlang A, Cunha G, Bastos J, Serra J, Barbosa B, Barroso L. Reliability Metrics for Generation Planning and the Role of Regulation in the Energy Transition: Case Studies of Brazil and Mexico. DOI: 10.3390/en14217428. Available from: https://www.mdpi.com/1996-1073/14/21/7428

[20] Billinton R, Allan R. Reliability Evaluation of Engineering Systems: Concepts and Techniques. 2nd ed. New York, NY, USA: Plenum Press; 1992

[21] Nooij M, Koopmans C, Bijvoet C. The value of supply security: The costs of power interruptions: Economic input for damage reduction and investment in networks. Energy Economics. 2007; **29**(2):277-295. DOI: 10.1016/j.eneco.2006.05.022

[22] Oren SS. Capacity payments and supply adequacy in a competitive electricity market. VII SEPOPE, Curitiba-Parana Brazil. 2000;**21-26**

[23] Crespo L. STE plants: Beyond dispatchability firmness of supply and integration with VRE. Energy Procedia. 2015;**69**:1241-1248

[24] Garrido PR. Effective load carrying capability (ELCC). PJM. 2020. Available from: https://www.pjm.com/-/media/committees-groups/task-forces/ccstf/2020/20200407/20200407-item-04-effective-load-carrying-capability.ashx

[25] Milligan M, Porter K. Determining the capacity value of wind: An updated survey of methods and implementation EXeter: NREL. Available from: https://www.nrel.gov/docs/fy08osti/43433.pdf

[26] Antuko. Capacity Balance Market a Plunge to Zero in 2020. Available from: https://antuko.com/capacity-balance-market-a-plunge-to-zero-in-2020/

Chapter 4

Management Methods of Energy Consumption Parameters Using IoT and Big Data

Carlos Daniel Valencia Rincón, Daniel Revelo Alvarado and Fernando Vélez Varela

Abstract

The continuous monitoring of electrical consumption helps to understand energy expenditure in functional consumption environments, such as campuses. For this reason, this work details the development of a mechanism that can do so, such as a network of sensors that is available in a telemetry system, which is determined to perform the acquisition and analysis of energy parameters. These actions are based on the concepts of Internet of Things (IoT) and Big Data. The acquired data are sent in a virtual local area network (VLAN), which is connected to a database server located in the campus environment, using the IoT concept, through the IEEE802.11/IEEE802.3 standards, so that later the analysis and monitoring of the electrical network can be carried out. For the construction of this prototype, noninvasive current sensors connected to a three-phase meter and a communication card are used to extract data from the meter and send it to the database. In the results, the possibility of specifying 30 energy parameters is obtained, with a packet loss rate equal to zero. With this network of sensors, whatever is in operation, such as low-voltage electrical power transformers, distribution boards, among others, can become intelligent data collection devices, from which information is extracted in real time by telemetry.

Keywords: power consumption, energy, Internet of Things (IoT), big data, telemetry, sensors, convergence

1. Introduction

The consideration of the impact of the use of electricity on the progress and economic evolution of a region is decisive and crucial. In this way, in economic costs, energy efficiency strategies and policies must also be formulated to accompany and help reduce electricity consumption [1]. When it is possible and there is the possibility, electrical transformers are observed by complex systems, which means high economic cost or, on the contrary, not being monitored, and there is no local or remote management of the reference variables defined in terms of the behavior of energy consumption [2]. Likewise, the term of electrical energy quality is an important aspect in what refers to the use of the energy supply [2]. In addition, the determination of the quality of electrical energy is defined by a wide variety of phenomena in electromagnetic physics, which are characterized by certain criteria, such as voltage and intensity, at a certain time [2]. Among the primary parameters of power quality, there is active power, reactive power, power factor, and voltage unbalance, among others [3].

Within the campus-type architectures, measurement devices can be arranged, which configure, know, and control energy consumption. The information obtained is useful when what is observed by the behavior of electricity consumption can be captured and processed as information, which must be directed to a management and control center, which can make the data collected available through an information system and visualization for both consumers and providers [1, 4, 5].

On the other hand, ICTs have offered innovative products and services that constantly change people's way of life [6]. Similarly, the Internet of Things (IoT) has undoubtedly opened a great opportunity, and that is to take advantage of the flexibility and efficiency of digital technology in daily life [4]. In addition, it mixes hardware and software technologies, communication protocols, and different processing technologies [6]. Likewise, it carries out a series of interconnections of equipment, personnel, processes, and data to achieve mutual communication and avoid problems. In this way, IoT can help improve different processes and make them more quantifiable and measurable through the collection and analysis of large amounts of data [6].

On the other hand, the presence of sensors as part of telemetry networks is one of the specifications that drives the development of the IoT in a keyway. These are used to collect and transmit data in real time, thus improving the efficiency and functionality of these IoT environments [7]. For the energy sector, there are four main pillars of IoT in energy saving: the transparency between assets, the monitoring and control of energy consumption at an elemental level, the optimization of energy consumption in real time, and the complete optimization of the system [4]. Likewise, it also includes the production, transmission, and distribution of energy, aspects in which use is made of a great variety of sensors, with which measurement processes are made, and it is through the management of the result of these that it could be possible to decrease in costs as in energy [7]. Furthermore, to improve energy efficiency, it is considered that the share of renewable energy must be increased, and the environmental impacts of energy use must be reduced [7]. In this way, IoT can help the energy sector transform from a centralized energy system to a distributed one with the integration of sensors with connectivity to this technology, that is, managed as a smart grid [3].

On the other hand, it is currently very pertinent to consider the use of IoT and Big Data platforms for the proper processing, management, and analysis of large amounts of data [8]. Likewise, to guarantee that users do not feel overwhelmed by the volumes of information, systems are required that can manage, analyze, and convert said data structure, which is obtained by dynamically processing and extracting an observed system [9]. In addition, in campus-type places, you can have a considerable number of transformers or electrical panels, which, with the integration of sensors, can obtain energy indicators. In this way, it is necessary to use technology, such as IoT and Big Data, among others, which can naturally face these challenges that exist today and, likewise, detect faults that could arise during the transmission or distribution of electricity [5].

The objective is framed in proposing a solution based on IoT and Big Data for the acquisition and analysis of energy parameters through a network of sensors using the IEE.802.11 protocol or the IEE.802.3 protocol. The final sensor network has two measurement devices, which manage to capture 30 variables; voltages, currents, powers, and frequency are among them. In addition, there is a dashboard in which the variables mentioned above can be seen in real time.

2. Background

Electricity demand has increased recently, and with the advancement of technology, many impulsive and nonlinear loads have been widely used in distribution networks. Similarly, the accuracy of energy metering is an important basis for the normal operation of the power grid. At present, not only power supply enterprises attach great importance to the development of electric power measurement technology, but also power users, large factories, and enterprises attach great importance to its development [9]. For example, Neve et al. [4] designed an IoT energy monitoring system, which consists of a PZEM-004 T-module-integrated thermocouple (TC) sensor, SD3004 electric energy measurement chip, and a Wemos D1 ESP8266 mini microcontroller for communication. The measurement data are sent to the database server via MQTT. As a result, they generated graphs in Grafana (dashboard) of voltage, current, and power of an electric furnace to demonstrate the operation. Likewise, Dharfizi [10] designed an IoT-based real-time energy monitoring system. It is composed of Raspberry Pi, which uses RS485 communication and node. js programming language to collect data from industrial energy meters (Schneider EM6400NG and Elmeasure EN8400) existing in the study company and store them locally. Therefore, they used Grafana (dashboard) to generate graphs showing the measured values of voltage, current, power, generated consumption, and current harmonics. Similarly, Tahiliani and Dizalwar [11] designed and implemented a minimalistic smart energy meter consisting of a microcontroller, an ACS712 current sensor, a Wi-Fi module, and an organic light-emitting diode (OLED) display. As a result, they generated graphs of nominal power and power consumption of some household appliances on the ThingSpeak platform, which is limited to connect multiple channels in its free version. In the work presented by them, a multi-interface smart energy meter communication following an IoT approach was proposed and demonstrated [2]. The electric meter is composed of an MSP430 processor, which reads parameters from the power grid, collects data such as voltage and current, and sends them to TM4C (processor) through UART (microcontroller). TM4C (processor) sends the data to nodeMCU through UART (microcontroller), which can perform post MQTT. Finally, the voltage, current, and power data are displayed in bar graphs. As a limitation, the device has internal memory capacity to store consumption data for the last 60 days of operation [12, 13].

In the work developed and specified in the article by Neve et al. [4], it is seen that they developed a module for measuring electrical variables in power transformers using IoT concepts. They used as materials a three-phase meter with the MODBUS RTU communication module, the MODBUS RTU to Transmission Control Protocol/Internet Protocol (TCP/IP) converter, a development board with the IEEE 802.11 module, and noninvasive current sensors. Likewise, to load the related information to the database (NoSQL), it was used the MQTT protocol. As a result, the voltage and current measurements in the three phases are presented graphically, in addition to the measurements of active power, apparent power, reactive power, and the average power factor. All these data were captured in a time interval of approximately 2 hours with a time loop of 5 seconds [12, 13].

Some of the work presented was applied to verify point consumption, which, unlike the work carried out, is focused on macro consumption measurements in campus-type sites.

3. Materials and methods/methodology

3.1 Methodology

To carry out the project, the methodology was divided into the following phases:

Phase I. Requirements analysis: The location of the electrical system in different areas of the campus environment was considered to measure the intensity levels of the wireless local area network (WLAN) systems.

Phase II. Logical design of the solution: Once the measurements of the WLAN systems were carried out, the optimal solution was sought for the acquisition and collection of data from the university campus cabinets.

Phase III. Development of the solution model: The data obtained were recorded in a storage infrastructure (database) on a server locally using the local area network or the IEEE 802.11 network. In addition, Big Data was used to process large amounts of data.

Phase IV. Test and optimization of the resulting model: Real-time data reading and storage tests were developed. In addition, connection tests were carried out between the database and the interface, and results were taken, which were used to optimize the final model.

Phase V. Implementation and testing of the resulting system: The implementation of the system was carried out in the different intelligent sensors that are in the electrical power supply areas of the campus air conditioners. In relation to the above, it was intended to take simultaneous samples to determine and guarantee the operation of the system.

Phase VI. Observation and interpretation of system variables: Observation and interpretation actions were carried out in the visualization interface from the tests of the system implemented in the digital power meters, and final samples were taken.

The system is generally arranged to continuously develop and test the telemetry processes, during certain periods, in the development environment determined for your application; generally, this is defined in a development life cycle of the test process, which can be composed of a set of stages that are predisposed, in which the limits must be appropriate for the correct collection of results [14, 15]. In the current situation, the research and implementation work process focuses on the phases and the development of a telemetry system through the acquisition and analysis of energy parameters based on IoT and Big Data concepts, providing it with a method according to the management of devices that become part of a converged system. For this reason, in addition to the corresponding analysis, it is also recommended to have a test bench for experimental practice.

3.2 Materials and procedures for development

The design of the sensor network of a telemetry system based on the concept of IoT and Big Data for the acquisition of energy parameters is built by hardware and software elements; these elements are presented in the following.

3.2.1 Hardware

One of the main characteristics of the creation of the IoT prototype that make up the sensor network is that it can be used as a final product and, in turn, is easy to move. Likewise, it is made up of the following materials:

- Three-phase meter with MODBUS RTU communication module.
- MODBUS RTU to TCP/IP converter.
- NodeMCU development board.
- Noninvasive current sensors.
- Breakers.
- GX16 connectors.

Source: Adapted from [4].

Table 1 presents the characteristics of the meter arranged for the integration of the IoT measurement sensor.

Table 2 presents the technical specifications of the Modbus RTU to TCP/IP converter for the integration of the IoT measurement sensor.

Table 3 presents the NodeMCU development board features for IoT measurement sensor integration.

3.2.1.1 Measurement device design

The development of the IoT sensor for three-phase measurement consists of the devices mentioned in **Table 1**, and it is corresponding to the hardware section. **Figure 1** shows initially a 3D model of the IoT sensor design with the devices that

Main technical parameters		
Voltage	Measuring range	380 V/100 V
	Consumo de energía	<1 VA
Current	Measuring range	5 A/1 A
	Energy consumption	<1 VA
Measure class	Active energy	0,5S
	Reactive energy	1S
Power supply	Measuring range	AC/DC 85–265 V
	Energy consumption	< 5 VA
Work temperature	Working range	−10–55°C
Frequency	Measuring range	45–65 Hz
Measurement error margin		±5%

Source: [16].

Table 1.
Technical references of the measurement module.

Main technical parameters	
Feeding range	5–36 VDC
Standard interfaces	RS232: 300–460.8 kbps RS485: 300–230.4 kbps
Network mode	Station/AP/AP + Station
Working mode	Transparent transmission/HTTP Client-Modbus TCP or Modbus RTU
Networking protocol	TCP/UDP/ARP/ICMP/DHCP/DNS/HTTP
Communication	Serial to Wi-Fi or Serial to Ethernet or Wi-Fi to Ethernet
Work temperature	−40–85°C

Source: [17].

Table 2.
Modbus RTU to TCP/IP converter technical references.

Main technical parameters	
Microcontroller	Tensilica Xtensa LX106 (32 BIT)
Interfaces	1 UART, 16 PWM (10 BIT), 2 SPI, 1 I2C, 1 I2S 1 ADC (10 BIT)
Flash memory	1 MB
SRAM	64 kB
Speed (MHz)	160
Wi-Fi	IEEE 802.11 b/g/n
Bluetooth	NO
Supply voltage	5 VDC

Source: [18].

Table 3.
Microcontroller technical references.

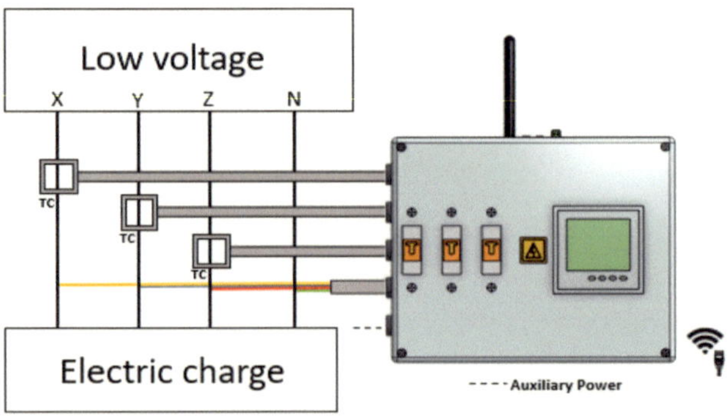

Figure 1.
3D modeling of the IoT sensor design and corresponding low-voltage installation. Source: prepared by the authors.

integrate it and its installation in lowvoltage. Finally, **Figure 2** shows a functional IoT device in a distribution board, connected to the current transformers of each phase through polyvinyl chloride cables (PVC) [19, 20].

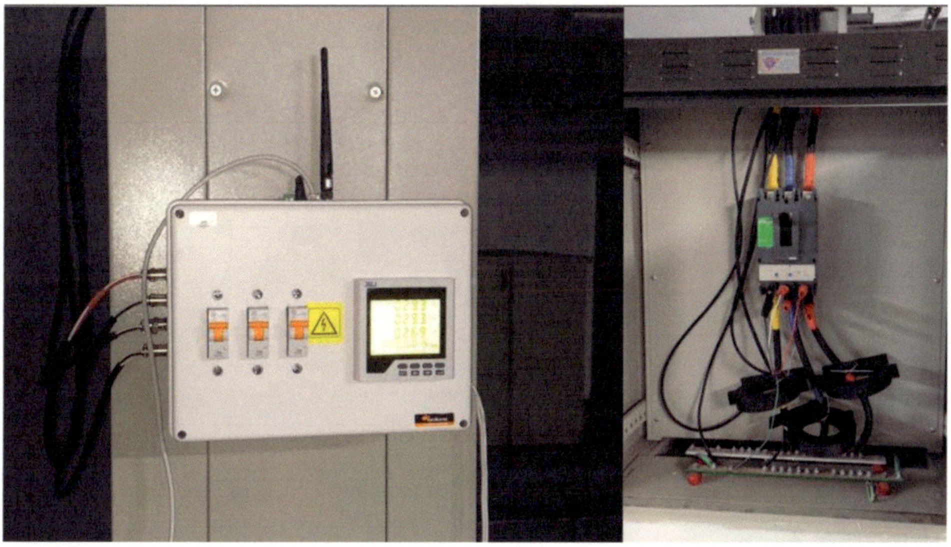

Figure 2.
Physical and functional design of the IoT sensor in a distribution board. Source: prepared by the authors.

3.2.2 Software

3.2.2.1 Network simulator

To have a first approximation to the network architecture, the modeling of the IoT sensor network was carried out, in which the power supply elements of the university campus are located using an appropriate network simulator for the case [14, 15], that is, the simulator with the required characteristics and specifications. In addition, a port-based virtual local area network (VLAN) was developed to have easy control of network traffic and better security of the data that are sent from the IoT sensor to the database server. **Figure 3** illustrates the scope that can be given to the project, the possible installation of 12 electrical network analyzers, each with its network interface that allows connection through the IEEE802.3 or IEEE802.11 communication protocol, and with capacity for greater installation, interconnected with each other in a campus-type unit. The yellow box represents the database server interconnected to a computer for viewing the data recorded by the IoT sensors from the energy distribution sites. The blue box shows the routing of one of the devices using the IEEE802.3 network, and the green box shows the routing of one of the devices using the IEEE802.11 network. The user can determine what type of connection to the communication network he wants. **Figure 4** shows the data collection from the server sent by the IoT sensor, thus giving the user the possibility to review the energy parameter collected separately and in real time. This information is being extracted from each of the points where the energy network analyzers are located and is subsequently being stored in a database dedicated solely to this information [21, 22].

3.2.2.2 IoT sensor communication mechanism

1. Communication with typical systems currently applied for online electrical measurements using the RS485 interface, two-wire full duplex (with Modbus protocol).

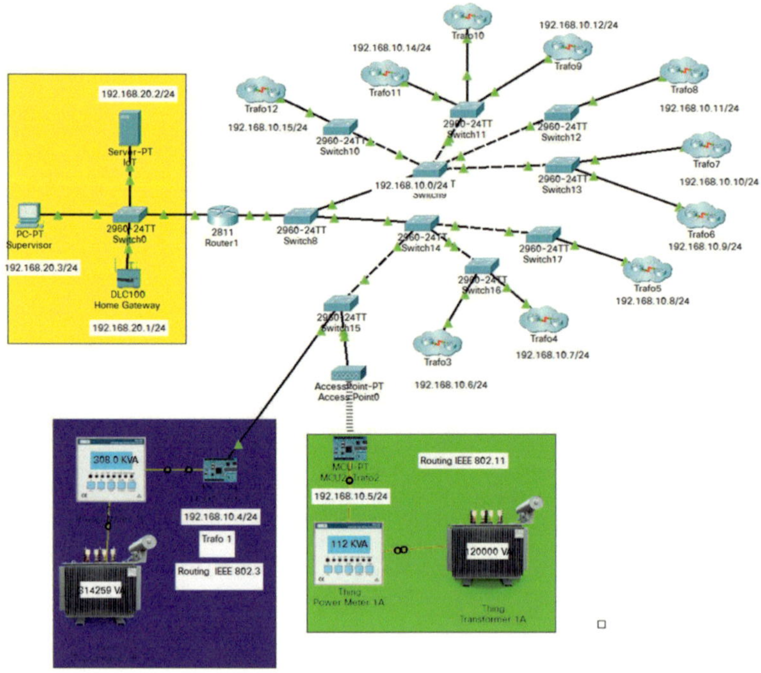

Figure 3.
Visualization of the final model of the network of electrical network analyzers of the university campus. Source: prepared by the authors.

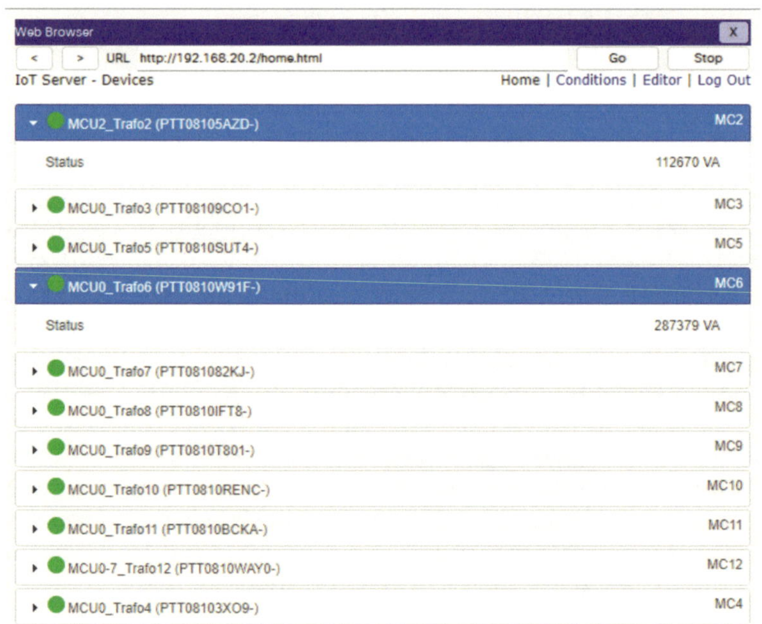

Figure 4.
Visualization of energy parameters distributed individually in a dashboard and previously stored in a database. Source: prepared by the authors.

2. Communication with typical external current sensors via portable multiconductor flexible cables composed of two soft copper conductors with individual thermoplastic polyvinyl chloride (PVC) insulation.

3. The meter's communication with the IEEE 802.11 or IEEE 802.3 system was done using a MODBUS RTU to TCP/IP converter, using the said converter in the slave mode.

4. Communication with the database was completed through NodeMCU with the IEEE 802.11 module and configured as the server mode; it was with the responsibility of consulting the meter and sending the received data to the database.

5. To upload the information to the database, the MQTT protocol was used.

6. The database is managed by a timing system that provides stable write speed and higher read speed (TSDB).

7. To view the energy parameters stored in the database, a dashboard made up of dynamic panels was used.

Source: Adapted from [4].

All IoT sensors communicated through a virtual local area network (VLAN) to increase productivity and data security and did not interact with other devices on another network and vice versa [21–23].

3.2.2.3 Data reception

Three-phase meters with RS-485 communication, also known as EIA/TIA-485, were used to receive data of all the energy parameters that are directed to the database server, each one connected to its respective voltage and current signals from of electrical distribution cabinets. The information that is obtained, is passed to an information system, that uses TCP/IP transmission protocol like mechanism of network, and using the Node-RED programming tool, that converts data from character to numeric format, as well as it connects all devices at same time, and manages the messages received [21–23]. The information storage was based on a time-series model, making use of the TSDB database, a database focused on IoT solutions, Big Data, and other technologies that collect a lot of information over time. The general architecture of the measurement system that outlines the communication of the sensor network is shown in **Figure 5**.

3.3 Materials and procedures for development

Using the data obtained from each meter as information, it is necessary to consider other electrical characteristics to help complement the information captured to give a correct interpretation. The following formula was used.

3.3.1 Offset angle

The phase angle is the representation of the power factor angle; it is obtained by finding the arc cosine of the power factor, as represented the following Equation [4]:

$$\varphi = \cos^{-1}\left(Pf\right). \tag{1}$$

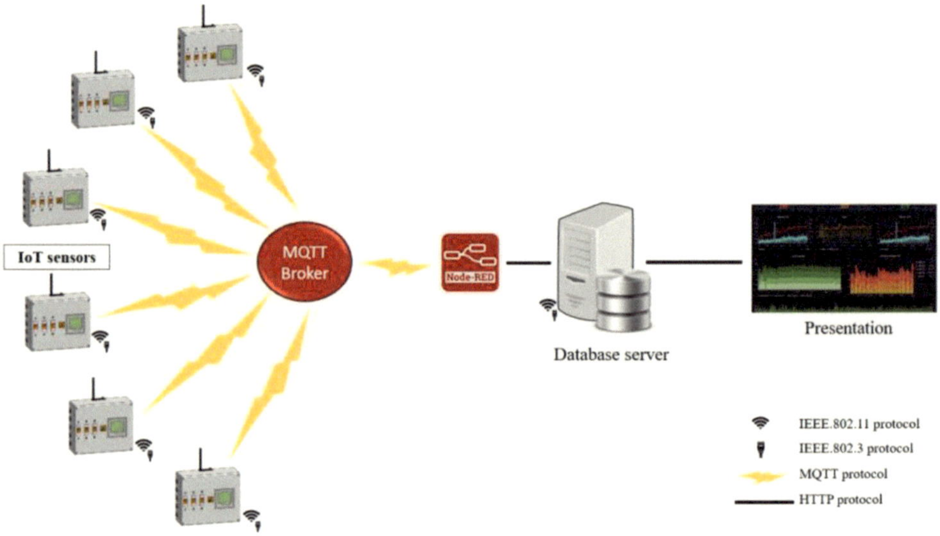

Figure 5.
General architecture of the sensor network. Source: prepared by the authors.

4. Results and discussion

As the main goal for the development of the project, it is proposed to carry out a network of sensors of a telemetry system through the acquisition and analysis of energy parameters based on IoT and Big Data concepts with a proposed solution, which was the electrical distribution cabinets or the power transformers of the university campus, which was determined using the concepts allusive to the management of power and electrical loads focusing on consumption. This section presents the results obtained once the operation process was carried out, together with the results obtained in the field tests, starting from the creation of a network of sensors integrating two electrical distribution panels to which the network analyzer was adapted to them, and through this, they become an intelligent object or smart object.

To make measurements in systems that reflect aspects, such as electrical power and what it represents, it is necessary to look for methodologies and models that, using technology and additional mechanisms, carry out such actions directly or indirectly. It is not uncommon that, today, millions of devices already bring with them a form of data extraction making use of protocol concepts with industrial-type determinations, which makes access to their information very limited, and it is not allowed to manage and analyze them. Likewise, it is not possible to connect a large number of said devices and make them converge between them, omitting the possibility of creating a sensor data collection network. Based on the above, we start from the search or creation of a mechanism that is capable of converting devices that do not have a form of data extraction to a device that does, making use of noninvasive external devices and determining it, as well as an IoT object, looking for the possibility of measuring variables or reference data of its operation in order to store and process said information in the future. On the other hand, it is also important to emphasize that using these mechanisms, it is possible to form an information network making use of telemetry processes collecting a large amount of data and making the devices converge in a database, and subsequently, the analysis can be carried out in this case of the fluctuations of the electrical network.

As already mentioned, the implementation characteristics used in this development are determined using simple concepts applied in the power measurement process, initially defined in two of the electrical distribution boards, which are derived from the campus transformers. For this, a network analyzer is used, which adapts to the necessary conditions for the collection of information. This device allows one to manage the data collected by making use of the network interface; it complements the network, which makes it an intelligent object.

The IoT sensor network in conjunction with the data acquisition system was used to monitor two electrical distribution cabinets on the university campus. The correct operation of this gives the possibility of collecting 31 energy parameters; likewise, validation tests of the different parameters obtained were carried out. The measurement network is fully functional, and each IoT sensor has a different topic as a unique identification for each energy parameter. The data collected are available on the university campus database server, where the different data generated by the sensors can be analyzed and visualized in real time. The visualization of these energy parameters is done using a free-to-use dashboard (open source) called Grafana. Abbreviations for the energy parameters are presented in **Table 4**

Table 5 presents the data obtained from the voltage of the three phases A, B, and C of the two measurement points located on the university campus during a period of

Abbreviation	Description
V_A	A phase voltage
V_B	B phase voltage
V_C	C phase voltage
V_{AB}	AB phase voltage
V_{BC}	BC phase voltage
V_{CA}	CA phase voltage
I_A	A phase current
I_B	B phase current
I_C	C phase current
P_A	A phase active power
P_B	B phase active power
P_C	C phase active power
P_S	Total active power
Q_A	A phase reactive power
Q_B	B phase reactive power
Q_C	C phase reactive power
Q_S	Total reactive power
Pf_A	A phase power factor
PF_B	B phase power factor
PF_C	C phase power factor
PF_S	Total power factor
S_A	A phase apparent power

Abbreviation	Description
S_B	B phase apparent power
S_C	C phase apparent power
S_S	Total apparent power
Phi $_A$	A phase Φ
Phi $_B$	B phase Φ
Phi $_C$	C phase Φ
Phi $_S$	Total Φ
F	Frequency

Source: prepared by the authors.

Table 4.
Abbreviation of energy parameters.

	Minimum	Maximum	Average
V_{A1}	8.1 V	128 V	97.6 V
V_{B1}	8.1 V	128 V	124 V
V_{C1}	8.1 V	128 V	122 V
V_{AB1}	216 V	223 V	220 V
V_{BC1}	216 V	220 V	219 V
V_{CA1}	217 V	223 V	220 V
V_{A12}	65.5 V	128 V	126 V
V_{B12}	114 V	128 V	126 V
V_{C12}	8.1 V	128 V	118 V
V_{AB12}	215 V	221 V	219 V
V_{BC12}	215 V	222 V	219 V
V_{CA12}	215 V	222 V	219 V

Source: prepared by the authors.

Table 5.
Values of the voltages obtained by the sensors installed in the distribution boards.

4 days; however, the measurement system can work constantly until it is determined that the information obtained is relevant. The acronyms of V_{A1}, V_{B1}, and V_{C1} represent the first measurement point, which is in block 6, floor 3 of the university campus. On the other hand, the initials V_{A12}, V_{B12}, and V_{C12} represent the second measurement point, which is in the basement of block 6. The values contained in **Table 5** are the maximum, minimum, and average value. Likewise, **Figure 6** shows the data collected and displayed on the graphical interface (dashboard) in a section of time of 5 minutes.

Figure 6 shows an interactive graph of three-phase voltage measurements on a university campus network at the installation site, where each line and its respective color have their specific abbreviation at the top right of the figure. Likewise, network fluctuations are observed in the measurements of both the sensors.

Similarly, **Table 6** shows the values of maximum, minimum, and the average of the normalized currents from 0 to 5 A of phases A, B, and C of the two measurement

DOI: http://dx.doi.org/10.5772/intechopen.105522

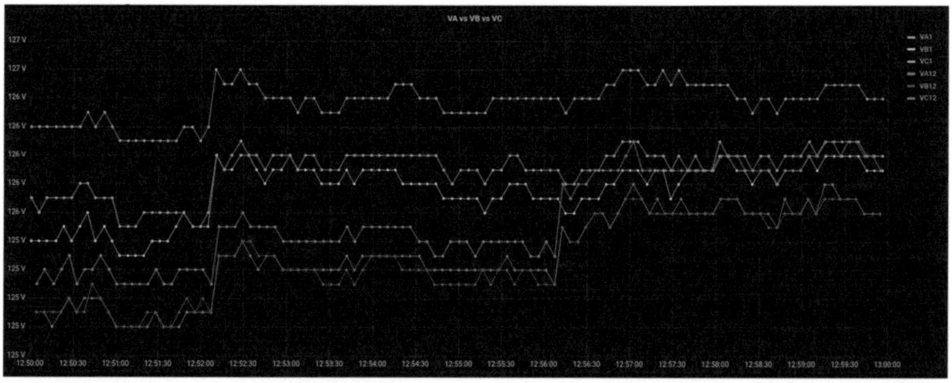

Figure 6.
Graphic interface of the visualization of the voltage energy parameters of both the sensors. Source: prepared by the authors.

	Minimum	Maximum	Average
I_{A1}	0.06 A	0.21 A	0.08 A
I_{B1}	0.02 A	0.22 A	0.04 A
I_{C1}	0.02 A	0.31 A	0.09 A
I_{A12}	0 A	0.44 A	0.05 A
I_{B12}	0 A	0.47 A	0.05 A
I_{C12}	0 A	0.41 A	0.03 A

Source: prepared by the authors.

Table 6.
Electric current values obtained by the measurement sensors.

points; **Figure 7** shows an interaction graph of the three-phase current measurements recorded by the sensors during a time window of 5 minutes, where each color of the signal is associated with an abbreviation located in the upper right part of the image. The IoT sensor located in the basement counts current transformers with a ratio of 1000 A to 5 A. In addition, the IoT sensor located on floor 3 of the same block has current transformers with a 600 A to 5 A ratio.

On the other hand, **Table 7** shows the active powers with their maximum, minimum, and average value of phases A, B, and C of the two measurement points. **Figure 8** shows the three-phase active power consumption levels recorded by the sensors in a time section of 5 minutes; in addition, during the measurements, it was observed that the active power consumption is high when the university campus is in academic activities.

On the other hand, when reviewing the results presented in **Tables 6** and **7**, it can be seen that the energy consumption in some buildings and in a place like in the basement of block 6 is higher than the consumption in floor 3 of the same block. Likewise, it can be seen that consumption on floor 3 is constant; therefore, its average is higher compared to that of the basement. This information can also be seen in **Figures 7** and **8**, which correspond to the currents and active powers of the sensors in a 5-minute time section.

Likewise, **Table 8** shows the reactive powers with their maximum, minimum, and average values of phases A, B, and C of the two measurement points. **Figure 9** presents a 5-minute section of the reactive power obtained by the three-phase measurement IoT

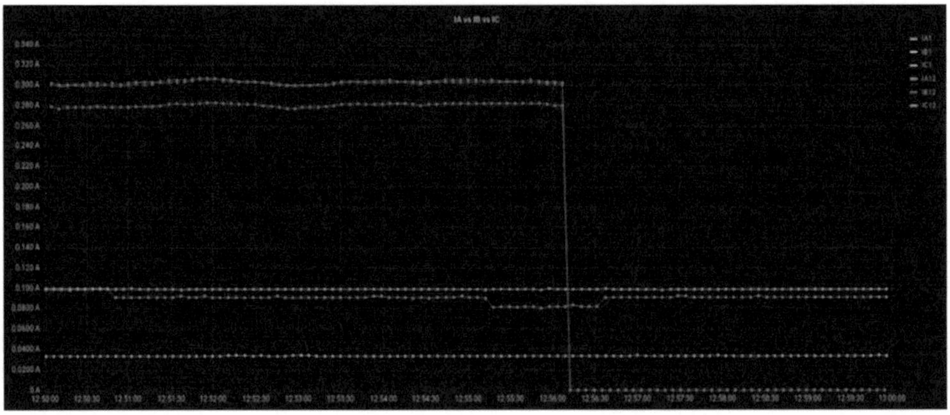

Figure 7.
Graphical interface for the visualization of the current variations of both the sensors. Source: prepared by the authors.

	Minimum	Maximum	Average
P_{A1}	1 W	56.4 W	26.8 W
P_{B1}	0.08 W	9.5 W	3.1 W
P_{C1}	3.5 W	36.8 W	20.2 W
P_{A12}	0 W	51.4 W	7.5 W
P_{B12}	0 W	44 W	6.2 W
P_{C12}	0 W	7 W	0.47 W

Source: prepared by the authors.

Table 7.
Values of the active powers obtained by the IoT sensors.

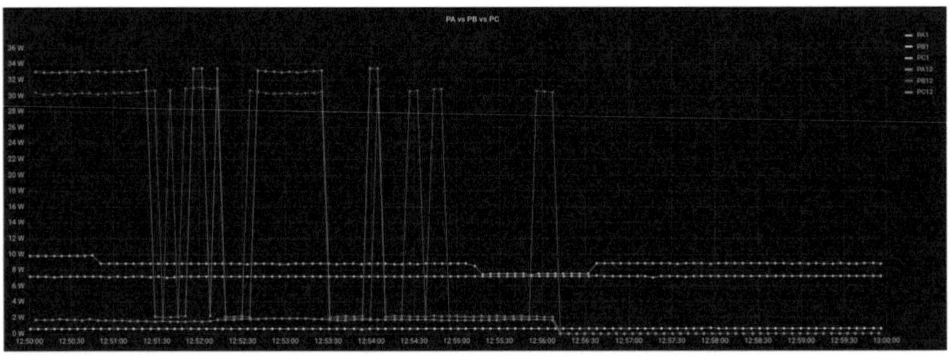

Figure 8.
Graphic interface for displaying the active power variations registered by each sensor. Source: prepared by the authors.

sensors during the entire measurement period; this power is related to the existence of coils or capacitors in the electrical installation associated with the distribution boards of the university campus.

	Minimum	Maximum	Average
Q_{A1}	0.37 VAR	54.6 VAR	26.9 VAR
Q_{B1}	0.51 VAR	24.3 VAR	11 VAR
Q_{C1}	0.74 VAR	53.2 VAR	31.1 VAR
Q_{A12}	0 VAR	24.1 VAR	3.37 VAR
Q_{B12}	0 VAR	33.2 VAR	5.18 VAR
Q_{C12}	0 VAR	47.5 VAR	7.60 VAR

Source: Prepared by the authors.

Table 8.
Values of the reactive powers obtained by the IoT sensors.

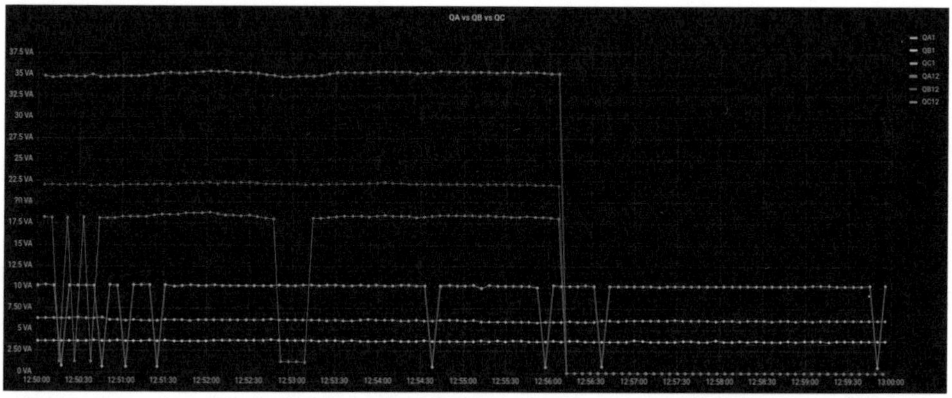

Figure 9.
Graphic interface for displaying the active power variations registered by each sensor. Source: prepared by the authors.

Similarly, **Table 9** shows the apparent powers with their maximum, minimum, and average values of phases A, B, and C of the two measurement points. **Figure 10** shows in a 5-minute section the interactions of the three-phase measurements of apparent power obtained by the sensors during the entire measurement period; this power is the sum of the energy that transforms these circuits in the form of heat.

	Minimum	Maximum	Average
S_{A1}	0.72 VA	79.3 VA	38.6 VA
S_{B1}	0.34 VA	27.8 VA	13.2 VA
S_{C1}	3.13 VA	65.2 VA	37.5 VA
S_{A12}	0 VA	55 VA	8.62 VA
S_{B12}	0 VA	53.2 VA	8.20 VA
S_{C12}	0 VA	51.6 VA	7.13 VA

Source: prepared by the authors.

Table 9.
Apparent power values obtained by the IoT sensors.

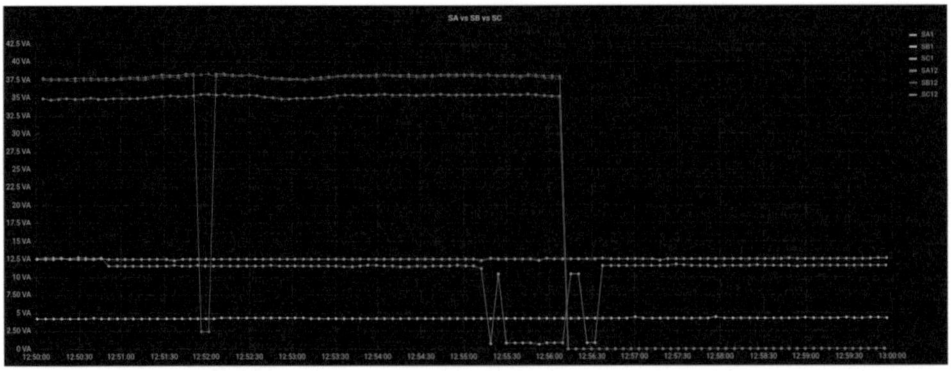

Figure 10.
Graphic interface for displaying the active power variations registered by each sensor. Source: prepared by the authors.

Table 8 corresponds to reactive power indicating that the sensor located on floor 3 of block 6 obtained higher reactance values compared to the meter located in the basement of the same block. In addition, **Table 9** shows the apparent power values where there was likewise greater apparent power in the electrical cabinet located on floor 3 of block 6.

Equally important, **Table 10** shows the power factors with their maximum, minimum, and average values of phases A, B, and C and the total of the two measurement points. Likewise, **Table 11** shows the phase angle of each meter.

Table 10 presents the power factor values for both the sensors; in the IoT sensor located in the basement of block 6 of the university campus, a maximum power factor value equal to one was presented; this is related to the time in which there was no electricity consumption. In the same way, from **Table 11**, it is possible to observe the values of the lag angles, giving an answer from another perspective that relates the apparent power and the active power, in other words, the lag in degrees existing between the intensity of the current and the voltage or voltage in the alternating current circuit.

The processing of energy indicators refers mainly to the support received from energy providers and the support and management of these parameters for the generation and transformation of electrical energy, as is the case of electrical powers,

	Minimum	Maximum	Average
Pfa_1	0.685	0.940	0.688
Pfb_1	0.0277	0.378	0.202
Pfc_1	0.04	0.604	0.527
Pfs_1	0.04	1	0.891
Pfa_{12}	0.087	1	0.97
Pfb_{12}	0.11	1	0.934
Pfc_{12}	0.037	1	0.823
Pfs_{12}	0.099	1	0.885

Source: prepared by the authors.

Table 10.
Values of the power factors obtained by the IoT sensors.

	Minimum	Maximum	Average
Phi $_{A1}$	45.76°	19.94°	46.52°
Phi $_{B1}$	88.41°	67.79°	78.35°
Phi $_{C1}$	87.71°	52.84°	58.19°
Phi $_{S1}$	87.71°	0°	27°
Phi$_{A12}$	85°	0°	14.07°
Phi $_{B12}$	83.7°	0°	20.93°
Phi$_{C12}$	87.88°	0°	34.61°
Phi $_{S12}$	84.26°	0°	27.75°

Source: Prepared by the authors.

Table 11.
Values of the offset angles.

	Minimum	Maximum	Average
P_{S1}	0.365 W	54.6 W	27.7 W
Q_{S1}	0.508 VAR	23.7 VAR	11 VAR
S_{S1}	0.742 VA	53.2 VA	31.7 VA
P_{S12}	0 W	24.1 W	3.3 W
Q_{S12}	0 VAR	33.2 VAR	4.68 VAR
S_{S12}	0 VA	47.5 VA	6.89 VA

Source: prepared by the authors.

Table 12.
Values of the total powers (P_S, Q_S, S_S) obtained by the IoT sensors.

which can be seen in **Table 12**, which represents the maximum, minimum, and average values of the total powers (active, reactive, and apparent). **Figure 11** shows the measurements obtained by the IoT sensors of the total powers (active, reactive, and apparent); these data are provided directly by the meters.

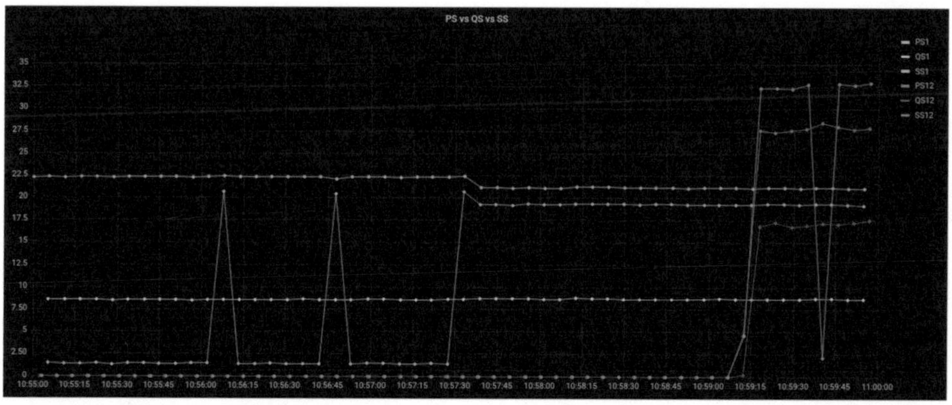

Figure 11.
Graphic interface for the visualization of the total power variations (active, reactive, and apparent) of the sensors. Source: prepared by the authors.

Number of measurements	Absolute time	Time	Source IP address	Destination IP address	Segmental latency	Total latency	Protocol	Size (Bytes)	Information publication message
280	30/11/2021 4:57	0.980271323	172.16.26.103	172.16.161.144	0	0.637971015	MQTT	77	va
284	30/11/2021 4:57	0.989446441	172.16.26.103	172.16.161.144	0.009175118		MQTT	193	vb, vc, vab, vbc, vca, ia
291	30/11/2021 4:57	1.011690827	172.16.26.103	172.16.161.144	0.022244386		MQTT	75	ib
299	30/11/2021 4:57	1.033896213	172.16.26.103	172.16.161.144	0.022205386		MQTT	75	ic
303	30/11/2021 4:57	1.056793208	172.16.26.103	172.16.161.144	0.022896995		MQTT	78	pa
306	30/11/2021 4:57	1.078475087	172.16.26.103	172.16.161.144	0.021681879		MQTT	77	pb
308	30/11/2021 4:57	1.101522884	172.16.26.103	172.16.161.144	0.023047797		MQTT	78	pc
311	30/11/2021 4:57	1.123173163	172.16.26.103	172.16.161.144	0.021650279		MQTT	78	ps
319	30/11/2021 4:57	1.156627794	172.16.26.103	172.16.161.144	0.033454631		MQTT	78	qa
321	30/11/2021 4:57	1.178479375	172.16.26.103	172.16.161.144	0.021851581		MQTT	77	qb
323	30/11/2021 4:57	1.200369057	172.16.26.103	172.16.161.144	0.021889682		MQTT	78	qc
336	30/11/2021 4:57	1.222415441	172.16.26.103	172.16.161.144	0.022046384		MQTT	78	qs
338	30/11/2021 4:57	1.245752241	172.16.26.103	172.16.161.144	0.0233368		MQTT	76	pfa
355	30/11/2021 4:57	1.268541235	172.16.26.103	172.16.161.144	0.022788994		MQTT	76	pfb
358	30/11/2021 4:57	1.291170826	172.16.26.103	172.16.161.144	0.022629591		MQTT	76	pfc
380	30/11/2021 4:57	1.313146809	172.16.26.103	172.16.161.144	0.021975983		MQTT	76	pfs
386	30/11/2021 4:57	1.346678341	172.16.26.103	172.16.161.144	0.033531532		MQTT	78	sa
404	30/11/2021 4:57	1.367970615	172.16.26.103	172.16.161.144	0.021292274		MQTT	77	sb
414	30/11/2021 4:57	1.390034699	172.16.26.103	172.16.161.144	0.022064084		MQTT	78	sc
417	30/11/2021 4:57	1.411740979	172.16.26.103	172.16.161.144	0.02170628		MQTT	78	ss

Number of measurements	Absolute time	Time	Source IP address	Destination IP address	Segmental latency	Total latency	Protocol	Size (Bytes)	Information publication message
428	30/11/2021 4:57	1.444022294	172.16.26.103	172.16.161.144	0.032281315		MQTT	75	f
437	30/11/2021 4:57	1.46624498	172.16.26.103	172.16.161.144	0.022222686		MQTT	78	vdesb
444	30/11/2021 4:57	1.487915959	172.16.26.103	172.16.161.144	0.021670979		MQTT	78	idesb
446	30/11/2021 4:57	1.520016373	172.16.26.103	172.16.161.144	0.032100414		MQTT	83	Angulo_fpa
450	30/11/2021 4:57	1.553387902	172.16.26.103	172.16.161.144	0.033371529		MQTT	83	Angulo_fpb
453	30/11/2021 4:57	1.58583512	172.16.26.103	172.16.161.144	0.032447218		MQTT	83	Angulo_fpc
456	30/11/2021 4:57	1.618242338	172.16.26.103	172.16.161.144	0.032407218		MQTT	83	Angulo_fps

Source: Prepared by the authors.

Table 13.
Segmentation of data transmission.

To evaluate the performance of the data transmission, the Wireshark program was used, which performs the analysis of protocols and data and allows visualizing the traffic that is happening on the network. The collection and sending of data from the IoT sensors, are scheduled every 5 seconds and are segmented into an average of 25 sections, as can be seen in **Table 13**, which shows the information obtained from the energetic measurements.

In addition, communication tests were carried out in sending and confirming packages for approximately 58.82 hours between the IoT sensors and the database server in a private network or VLAN, in order to verify the sending and receiving of data from where the results presented in **Tables 14** and **15** were obtained, which means that in the 58.82-hour interval of sending and receiving information, there was no loss or forwarding of packets, which provides good scalability and flexibility in the network.

The total information of the packet traffic in 1-second intervals during the measurement time had a duration of 58.82 hours; to visualize the information in a more detailed way, **Figure 12** shows the traffic analysis information in a window of duration of approximately 1.67 minutes; the time spaces between the peaks represent intervals when the IoT sensor is not communicating at all with the database server; for that reason, the packets delivered are 0. In addition, each signal peak refers to the number of packets that are sent and describes the number of sequences created during the transmission of information, that is, the number of measurements (280), (284) and (291) shown in **Table 13**, corresponds to eight of the thirty-one variables. In that case, the measurement the value two hundred and eighty (280), is the information

Parameters	Value
Total packages sent	994,618
Total packets received	994,618
Total discarded packages	0
Bytes sent from server to IoT sensor	77 MB
Bytes sent from IoT sensor to server	51 MB
Average latency	23.20 ms

Source: prepared by the authors.

Table 14.
Results of the data transmission performance of the meter block 6 third floor.

Parameters	Value
Total packages sent	883,251
Total packets received	883,251
Total discarded packages	0
Bytes sent from server to IoT sensor	71 MB
Bytes sent from IoT sensor to server	47 MB
Average latency	20.70 ms

Source: prepared by the authors.

Table 15.
Performance results of meter block 6 basement data transmission.

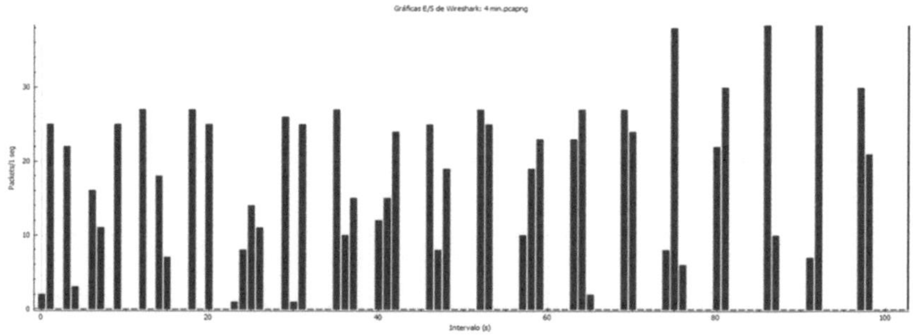

Figure 12.
Traffic analysis graph of the measurement for 58.82 hours approximately in an interval of 1.67 minutes. Source: prepared by the authors.

corresponding to the voltage in phase A that it was sent, this is the measurement corresponding to the first packet sent; in the measurement (284), it was sends the information contained from the voltage in phase B to the current in phase A; likewise, this is done with the measurement (291), this information is corresponding to the current in phase B is sent. It should be noted that the data frame is not it ends until the last measurement number packet (456) has been sent, which corresponds to the variable "angle_fps". Finally, the duration time of each data frame corresponds to an average of 0.63 seconds and is composed of the time intervals when the sensor sends information and when the sensor does not share information with the database server.

This is the mature concept of a useful telemetry mechanism suitable for an interpreter who is familiar with electrical power consumption mechanisms to be able to perform, analyze, and interpret the information collected alluding to energy consumption. Therefore, those who are interested in this part know that it must be conditioned or improved and, in that case, must regulate the consumption generation processes; with this, a certifiable concept is allowed and is also supported by the green consumption standards of energy and by the ISO50001 standards that currently govern in Colombia.

There are currently different types of electrical network analyzers on the market, for which costs vary depending on customer needs. The designed modules prototypes, convert power transformers and the derived electrical distribution boards, into a class of smart object, and this allows to generate a network of sensors of low cost and with a reliable and safe data transmission method, in other words, when these components are in operation, there is not will be loss of information. Likewise, the customer can determine the measurement time required for acceptable results, and which electrical variables they want to acquire and know, furthermore, and if they want to use the transmission method remotely using the IEEE802.11 standard, or locally using the IEEE802.3 standard. In addition, the end customer can view the energy parameters through a graphical interface (dashboard) in a dynamic and interactive way.

The sensor network during the development of this work has a database server hosted on the university campus, which has a wide storage limit, for the integration of more sensors and the massive collection of more energy parameters having remote access.

5. Conclusions

According to the results obtained with the operation of the sensor network based on the concept of IoT and Big Data, the following advantages could be evidenced:

- The method of data acquisition, transport, and processing of energy parameters used satisfactorily achieved acceptable performance, achieving zero loss of information packets as well as latency with permissible response time.

- The network of energy measurement sensors is convergent in all its measurement sensors, which indicates that the architecture of this network is functional.

- The database server developed for the storage of information supports the processes based on IoT and Big Data, which allows us to have a faster response time to visualize the energy parameters captured by the sensors based on time.

- The devices developed allow the user easy handling and installation. Likewise, it is multifunctional, which means that it can measure not only three-phase voltages but also single- and two-phase voltages.

- The results suggest that the structured sensor network does not require a large bandwidth for the transmission of information.

- The proposed sensor network is in the investigative exploration stage and, in turn, in this work, can serve as a guide for generating feedback and developing future projects based on this work.

- To advance the project, other communication mechanisms, such as low-frequency laser radio, must be considered, as this allows a greater range of connectivity and communicates devices, where the IEEE802.11 and IEEE802.3 standards are not present.

- The work carried out has two measurement devices; it is recommended in the future to carry out the measurement processes using four or more devices that converge in the proposed data acquisition system. Likewise, make a robust analysis of the data obtained using Big Data processes to be able to make predictions in the future.

Acknowledgements

Thanks to the Santiago de Cali University, to the COMBA I+D group, for the sponsorship and support for the development of this research work.

Author details

Carlos Daniel Valencia Rincón, Daniel Revelo Alvarado and Fernando Vélez Varela*
Faculty of Engineering, Program of Electronic Engineering, University Santiago de Cali, Valle del Cauca, Colombia

*Address all correspondence to: fernando.velez00@usc.edu.co

References

[1] Carrasco Chipantiza LF. Sistema de monitoreo y telegestión del consumo eléctrico en cargas residenciales basado en una arquitectura IoT [En línea]. Carrera de Ingeniería En Electrónica y Comunicaciones, Facultad de Ingeniería En Sistemas, Electrónica e Industrial, Universidad Técnica de Ambato; 2018. Available from: https://repositorio.uta.edu.ec/handle/123456789/28809

[2] Van Aubel PE. Smart metering in the Netherlands: What, how, and why. International Journal of Electrical Power & Energy Systems. 2019;**109**:719-725

[3] Collins Mendoza MJ, Lucas Villao II, Silva Becherán JM. Análisis de evento de calidad de energía en una red eléctrica de baja tensión mediante el analizador MPQ2000 Megger. Repositorio Institucional de La Universidad Politécnica Salesiana; 2020

[4] Neve VG, Pullawar PV, Kidile CH. Power quality issues & detection methods for voltage sag and swell. International Journal of Research in Biosciences, Agriculture and Technology. 2020;**II**(Special-16):71-79

[5] Avancini DB, Rodrigues JJPC, Martins SGB, Rabêlo RAL, Al-Muhtadi J, Solic P. Energy meters evolution in smart grids: A review. Journal of Cleaner Production. 2019;**217**:702-715

[6] Velez Varela F. In: Valle UD, editor. Sistema informático de gestión para la evaluación del impacto ambiental de la infraestructura de las tic, redes y servicios en una red LAN. Vol. 53. Santiago de Cali: Repositorio digital Universidad del Valle; 2016

[7] Ortiz JH, García JFC, Khalaf OI, Varela FV, Baron PJB, et al. Development of a module for measuring electrical variables in power transformers based in IoT, to manage and monitoring by telemetry mechanism. Journal on Internet of Things. 2021;**3**(2):53-63

[8] Shrouf F, Ordieres J, Miragliotta G. Smart factories in Industry 4.0: A review of the concept and of energy management approached in production based on the Internet of Things paradigm. In: IEEE International Conference on Industrial Engineering and Engineering Management. Vol. 1. 2014. pp. 697-701

[9] Yassine A, Singh S, Shamim Hossain M, Muhammad G. IoT big data analytics for smart homes with fog and cloud computing. Future Generation Computer Systems. 2019;**91**:563-573

[10] Dharfizi AD-H. The energy sector and the Internet of Things—Sustainable consumption and enhanced security through Industrial Revolution 4.0. Journal of International Studies. 2018;**14**:99-117

[11] Tahiliani VDM. Green IoT systems: An energy efficient perspective. In: 2018 11th International Conference on Contemporary Computing (IC3). 2018

[12] Motlagh NH, Mohammadrezaei M, Hunt J, Zakeri B. Internet of things (IoT) and the energy sector. Energies. 2020;**13**(2):1-27

[13] Terroso-Saenz F, González-Vidal A, Ramallo-González AP, Skarmeta AF. An open IoT platform for the management and analysis of energy data. Future Generation Computer Systems. 2019;**92**:1066-1079

[14] Xin Y, Mingshuai C, Xinyang L, Hongwei Z, Zong-Qiang S, Liwei L.

Research of three-phase high-voltage energy metering device. In: 2017 Chinese Automation Congress (CAC). 2017. pp. 5845-5848

[15] Muralidhara S, Hegde N, Rekha PM. An internet of things-based smart energy meter for monitoring device-level consumption of energy. Computers and Electrical Engineering. 2020;**87**:106772

[16] Chujing Electric. Shanghai Chujing Electric Co. Ltd. [En línea]. 2014. Available from: http://www.chujing-electric.com/English/About%20us/

[17] Jinan USR IOT Technology Limited. h.usriot.com [En línea]. 2021. Available from: https://en.usr.cn/

[18] Espressif Systems Solutions. Espressif Systems (688018.SH) [En línea]. 2022. Available: https://www.espressif.com/en

[19] Chooruang K, Meekul K. Design of an IoT energy monitoring system. In: 2018 16th International Conference on ICT and Knowledge Engineering (ICT&KE). 2018

[20] Mudaliar MD, Sivakumar N. IoT based real time energy monitoring system using Raspberry Pi. Internet of Things. 2020;**12**:100292

[21] Avancini DB, Rodrigues JJPC, Rabêlo RAL, Das AK, Kozlov S, Solic P. A new IoT-based smart energy meter for smart grids. International Journal of Energy Research. 2021;**45**(1):189-202

[22] Velez Varela F. Energy consumption model for green computing. In: Mobile Computing. London, UK: IntechOpen; 2020. pp. 1-14

[23] Bollen MHJ, Gu IYH. Signal Processing of Power Quality Disturbances. Wiley-IEEE Press; 2006

Section 2

Green Energy for Computing Networks

Chapter 5

From Photovoltaic to Agri-Natural-Voltaic (ANaV)

Giovanni Campeol, Lorella Biasio, Silvia Foffano, Davide Scarpa and Giulio Copparoni

Abstract

Italy is the geographic area with the highest world concentration of cultural sites, landscapes with a high esthetic value, and biodiversity. Therefore, any modification in the territory is to be performed by highly considering these cultural, landscape, and natural values. Also, Italy has a high human pressure on agricultural areas, especially in flatlands, that generally have a high agronomic value. As a consequence, the planning of photovoltaic installations in agricultural areas must meet at least five basic criteria: to cause as little use as possible to the agricultural soil, to maintain agricultural activities, to strengthen or introduce natural habitats, to properly mitigate the landscape impact, and to be located as far as possible from residential areas. This chapter presents the technological innovation with a high environmental value that characterizes the photovoltaic system, called Agro-Natural-Voltaic (AnaV), also using methodological and planning schemes together with landscape simulations. Moreover, the essay gives a description of the study on the environmental impact (for the administrative procedure of the Environmental Impact Assessment), laid down for the emblematic case study.

Keywords: renewable energies, photovoltaic, agriculture, biodiversity, landscape

1. Introduction

Italy is the geographic area with the highest world concentration of cultural sites, landscapes with very high aesthetical value, and biodiversity. It is also the country with the highest number (59) of sites on the UNESCO World Heritage List. In fact, new properties were added to the WHL: on 26 July 2021 "Padua's fourteenth-century fresco cycles and Montecatini Terme," and on 28 July the "porticoes of Bologna" and 8000 hectares of Italian forests within the transnational candidature "Primeval beech forests of the Carpathians and other Regions of Europe." In this special ranking, China is second with 56 sites and Germany is third with 50 sites.

They are prevalently cultural sites, some of which with considerable natural and landscape importance.

Therefore, any modification in the territory is to be performed by highly considering these cultural, landscape, and natural values.

The great promotion of renewable energy (especially sun and wind) has led to a conflict in Italy between those who want to preserve the landscapes and those who want to transform them regardless of the characteristics of renewable energy projects, which very often produce a heavy bureaucratic blockage.

Before transforming a territory, the project of renewable energy must forcibly be environmentally compatible with the site concerned and not vice versa, and this is of crucial importance in the geographic case of Italy.

2. Photovoltaic

There are several types of photovoltaic projects [1], as synthetically presented in **Table 1**.

Italy has greatly developed photovoltaic technology and, in this case, also its production modes. In particular, a different conceptual paradigm has been recently proposed: first, the environmental characteristics of a geographic area are studied, then, the kind of photovoltaic is defined.

The case study presented in this essay seems to be the most important "system technological innovation" that characterizes the planning of photovoltaic in Italy (as well as in all those geographic areas characterized by high environmental sensitivity).

Figure 1 summarizes the evaluation process realized for the photovoltaic project called "Agri-Natural-Voltaic (ANaV), proposed by Tozzi Green SpA located in Mezzano (Province of Ravenna) in Italy. The project was approved in 2022 by the National Committee of Environmental Impact Assessment of the Italian Ministry of Ecological Transition[1].

3. The "Agri-natural-voltaic" technology (ANaV)

Considering the analysis of the different projects for the photovoltaic system [2], presented above, the ANaV project is compared to other projects with the same technology but with different uses of the soil (called Scenarios).

The comparison is carried out through a comparative table (**Table 2**) to better understand the characteristics of the different photovoltaic projects, which shows the best environmental sustainability of the innovation of the ANaV project.

Preliminary environmental assessments have consequently led to optimize the ANaV photovoltaic technology, which is the most performing, in relation with the reference geographic context (flatlands in the municipality of Cerignola, Foggia Province in Puglia Region – Italy).

The geographic location and the characteristics of the ANaV project are represented in the following images (**Figures 2** and **3**).

The ANaV Project allows a suitable realization of the initiative in the environmental context, safeguarding agricultural production and, simultaneously, positively influencing the botanic, vegetational, and wildlife contexts of the area.

The visual characteristics of the ANaV project are represented in the following representations (**Figure 4**) [3].

---

[1] Complete title: *Progetto per la realizzazione dell'impianto (ANaV) per la produzione di energia elettrica da fonte solare della potenza complessiva di 99,42 MW, sito nel comune di Cerignola, località "San Giovanni in Fonte" e relative opere di connessione nei comuni di Stornarella, Orta e Stornara (FG) Italia.*

Type	Max installation height	Interaction with agriculture	Investment cost	Cost/maintenance	Installation efficiency
 Fixed mount solar panels	Max height of about 4 m	Poorly suitable due to excessive shading and difficult use of agricultural vehicles. 10% of the ground surface of the installation can be exploited	Reduced	Simple and cheap ordinary maintenance	Lower expected productibility
 Single-axis: roll tracker	Panels to max tilt about 4.5 m	Suitable for double-sided modules that reduce shading. 30% of the ground surface can be exploited	3–5% higher than the fixed mount system	Simple and cheap ordinary maintenance. Compared with the fixed mount system, additional costs for the maintenance of the track system motors	Compared with the fixed mount system, 15–18% higher productibility (following the latitude of the location)
 Single-axis: polar tracker	About 6 m high (the panel is tilted following the terrestrial axis)	Suitable for double-sided modules that reduce shading; a very complex structure with concrete plinths that hinder the use of agricultural vehicles	10–15% higher than the fixed mount system	Simple and cheap ordinary maintenance. Compared with the fixed mount system, additional costs for the maintenance of the tracking system motors	Compared with the fixed mount system, 20–23% higher productibility (following the latitude of the location)
 Single-axis: azimuth tracker	The structure can be up to 8–9 m high	The surrounding area must be free to ensure the panel rotation; therefore, it cannot be used for agriculture	25–30% higher than the fixed mount system	More expensive ordinary maintenance due to the installation height. Additional costs for the maintenance of the motor (for the rotation) and the cleaning of guideways	Compared with the fixed mount system, 20–22% higher productibility (following the latitude of the location)

Type	Max installation height	Interaction with agriculture	Investment cost	Cost/maintenance	Installation efficiency
 Dual-axis tracker	The structure can be up to 8–9 m high	Suitable for agriculture (also with automated vehicles). 30% of the ground surface of the installation can be exploited	25–30% higher than the fixed mount system	More expensive ordinary maintenance due to the installation height. Additional costs for the maintenance of the two motors of the dual-axis tracking system	Compared with the fixed mount system, 30–35% higher productibility (following the latitude of the location)
 Dual-axis tracker on an elevated structure	The structure can be up to 7–8 m high	Suitable for agricultural use (also with big automated vehicles). 70% of the ground surface of the installation can be exploited, with crops up to 3–4 m high	45–50% higher than the fixed mount system	More expensive ordinary maintenance due to the system height. Additional costs for the maintenance of the two motors of the dual-axis tracking system	Compared with the fixed mount system, 30–35% higher productibility (following the latitude of the location)

Table 1
Comparison between different types of photovoltaic projects.

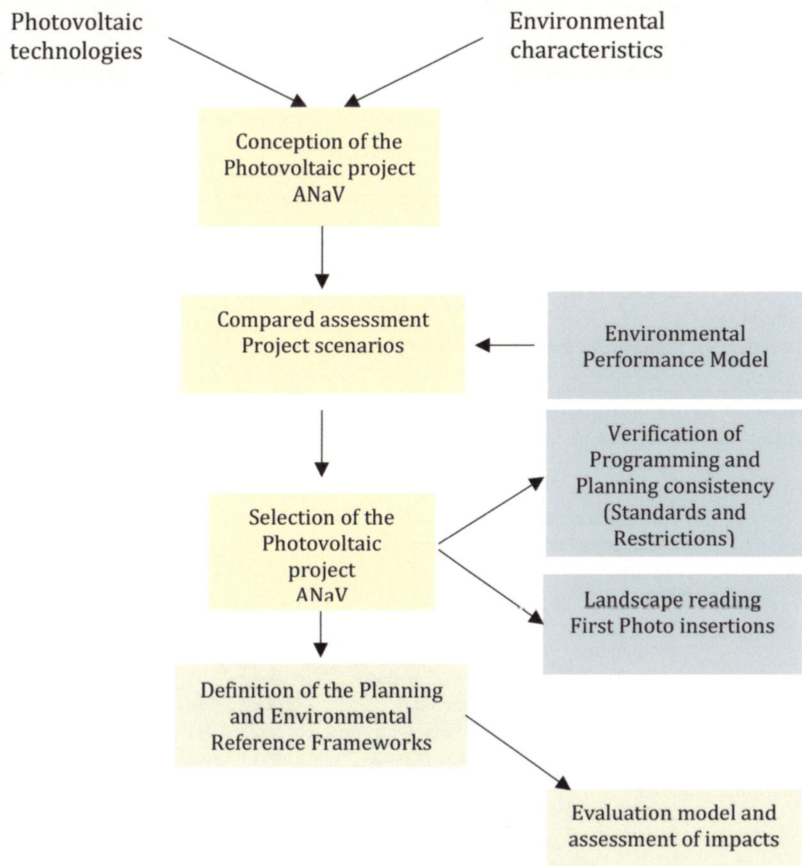

Figure 1.
Conceptual table: environmental feasibility and its relation with the environmental assessment.

The ANaV project is made up of modules on "single-axis trackers" for a total capacity of 99.42MWp, suitably elevated from the ground and positioned in such a way as to be best suited to agricultural activity on the same ground surface. (**Figure 5**).

The photovoltaic modules (double-sided with nominal unit power of 605 Wp) are 1.3 × 2.2 m in size, and 4 cm thick. They are in pairs, mounted horizontally with respect to the main axis of the tracker. Twenty-eight modules will be mounted on each single-axis tracker.

The special structure of photovoltaic panels planned for the ANaV system allows high flexibility in agricultural interventions both for machine accessibility and for the selection of crops and cultivation methods (**Figure 6**).

The panel height from the ground ensures the proper airing in the underlying area, favoring the normal growth of herbaceous vegetation and, contemporarily, preserving the normal indigenous microbial activity of the soil; also, rainwater can flow below without interfering with the normal underground drainage and storage.

In addition, the position of panels in parallel, equidistant rows, enables a rational organization of cultivations and rotations and/or cultivation changes.

The cultivations envisaged by the ANaV project [4] produce a high remuneration per hectare in front of high demands for labor. Their nature of root crops requires

	Scenario 1	Scenario 2	Scenario 3	Scenario 4 (ANaV)
	Type of system			
Project characteristics	**Fixed mount photovoltaic**	**Intensive agri-voltaic (single-axis trackers + super intensive olive growing system)**	**Diversified agri-voltaic (single-axis trackers + diversified agriculture)**	**Agri-natural-voltaic (ANaV) (single-axis trackers + diversified agriculture and renaturalization)**
Usable agricultural area	10% Net of the ground surface occupied by the installation	30% Net of the ground surface occupied by the system	76.35% Including the sub-panel belts	89.32% Including the sub-panel belts and the green belts
Possible cultivations	Artificial grassing (non-natural, non-spontaneous species). Working stages: 1. late spring/early summer—soil tillage, with mechanical weeding by agricultural tractors and inter-row milling machines to prevent fire risks related to spontaneous herbs drying up. 2. winter—sowing with conventional agricultural machinery. 3. early spring—grass shredding.	Field farming in spacings. Setting up of a row on a central axis along a string line with a stake at the beginning and at the end of the row for good anchoring. Non-cultivable areas: near the photovoltaic modules, with the max horizontal covering, artificial grassing (non-natural, non-spontaneous species) will be introduced. Grass-cutting could be unnecessary since agricultural crops could help clean the areas under the panels with ecologic hot-water weeding.	Field farming in spacings + lateral belts under the panel since the tilting movement enables land use for agricultural crops. The expected growing area is part of the existing biological agriculture and is organized in long-term (10 years) crop rotations. Adoption of agricultural crops in line with local crops: artichoke; asparagus; crop rotation of cereals and legumes; honey plants; non-cultivable areas near the photovoltaic modules, in a row of about 1 m around the panel support poles.	Field farming in spacings + lateral belts under the panel since the tilting movement enables land use for agricultural crops. The expected growing area is part of the existing biological agriculture and is organized in long-term (10 years) crop rotations. Adoption of agricultural crops in line with local farming: artichoke; asparagus; crop rotation of cereals and legumes; honey plants; noncultivable areas near the photovoltaic modules, in a row of about 1 m around the support poles.
Description of renaturalization operations and relevant management	No renaturalization interventions to stop spontaneous colonization	No renaturalization interventions to stop spontaneous colonization	No renaturalization interventions to stop spontaneous colonization	A wide area on the lateral belts of the system is envisaged for the development of the surrounding natural habitats 6220, arid Mediterranean meadows. This is a semi-natural environment, residual with respect to previous agricultural use or

	Scenario 1	Scenario 2	Scenario 3	Scenario 4 (ANaV)
	Type of system			
Project characteristics	**Fixed mount photovoltaic**	**Intensive agri-voltaic (single-axis trackers + super intensive olive growing system)**	**Diversified agri-voltaic (single-axis trackers + diversified agriculture)**	**Agri-natural-voltaic (ANaV) (single-axis trackers + diversified agriculture and renaturalization)**
				coming from sparse grazing. Its realization does not only integrate the installation area with the surrounding environmental mosaic, but it also helps local biodiversity and grazing/beekeeping
Increase of biodiversity	Increase of fodder plant species, limited by the perennial shading of some belts caused by fixed panels	Increase of fodder plant species. Partially limited by the perennial shading of some belts caused by fixed panels.	Increase because of crop diversification and fodder plant species. Minimally limited by the perennial shading of some belts near the structure mount.	Increase of biodiversity that maximizes the integration between agricultural and naturalistic features. Minimally limited by the perennial shading of some belts near the structure mounts. The agronomic module is part of a wider scenario integrated with the naturalistic part that is expected to use a wide surface on the lateral belts of the installation for the development of the area natural habitat (6220 – arid Mediterranean meadows)
Description of agricultural activities (use of chemicals, water consumption, etc.)	None	Installation of a system of drip irrigation. Phytosanitary treatments with a pneumatic nozzle that channels water jets only on one side, and is trained by a tractor.	Since the land concerned is presently managed in a certified biological regime, also in the future the organic directives will be valid, and therefore special attention	Since the land concerned is presently managed in the certified biological regime, also in the future organic directives will be followed, and therefore special attention

	Scenario 1	Scenario 2	Scenario 3	Scenario 4 (ANaV)
	Type of system			
Project characteristics	**Fixed mount photovoltaic**	**Intensive agri-voltaic (single-axis trackers + super intensive olive growing system)**	**Diversified agri-voltaic (single-axis trackers + diversified agriculture)**	**Agri-natural-voltaic (ANaV) (single-axis trackers + diversified agriculture and renaturalization)**
			will be paid on the organization of crop rotations. Together with the non-use of active substances (fertilizers or pesticides), these represent one of the basic principles of agricultural sustainability. Traditionally, irrigation is not planned (except in case of emergency and only for artichoke crops).	will be paid on the organization of crop rotations. Together with the non-use of active substances (fertilizers and pesticides), these represent one of the basic principles of agricultural sustainability. Traditionally, irrigation is not planned (except in case of emergency and only for artichoke crops).
Description of the project with reference to the landscape (structural reading)	Total, permanent, and standardizing modification of farmland, which is represented as a totally used productive area.	Partial and permanent modification of farm landscape, since part of the land is used for mechanized olive-growing.	Partial and dynamic modification of farm landscape, since free areas are cultivated in rotational crops.	Partial and dynamic modification of farm landscape, since free areas are cultivated in rotational crops. In line with the landscape that characterizes this agricultural section, the project includes olive groves, orchards, and vineyards in the perimeter belts of the ANaV installation, following the patterns and crops existing across the Provincial Road SP 83 ("tratturello," sheep track), and the olive groves of the agricultural landscape along the Provincial Road SP 95.
Description of the project with	Total obstruction of the perceptive	Total obstruction of the perceptive	Dynamic partial obstruction of the	Dynamic partial obstruction of the

	Scenario 1	Scenario 2	Scenario 3	Scenario 4 (ANaV)
	Type of system			
Project characteristics	**Fixed mount photovoltaic**	**Intensive agri-voltaic (single-axis trackers + super intensive olive growing system)**	**Diversified agri-voltaic (single-axis trackers + diversified agriculture)**	**Agri-natural-voltaic (ANaV) (single-axis trackers + diversified agriculture and renaturalization)**
reference to the landscape (perceptive reading)	vision due to the type of the system	vision due partly to the installation and partly to super-intensive olive growing	optical cones (longitudinal and orthogonal): longitudinal: limited optical obstruction since the panels do not interfere with the landscape. As regards the width of the optical cone, the view is open only with horizontal panels; orthogonal: variable according to the panel orientation	optical cones (longitudinal and orthogonal): longitudinal: limited optical obstruction since the panels do not interfere with the landscape. As regards the width of the optical cone, the view is open only with horizontal panels; orthogonal: variable according to the panel orientation. For consistently integrating with the landscape and reducing the view of the technological parts, the project envisages to introduce orchards, olive groves, and vineyards in the perimeter belts.
Farming profitability	Grass cover is not significantly profitable.	The super-intensive can dramatically reduce the need for labor, not only for harvesting operations (in the traditional system this means up to 80% of the total costs) but also for all the other mechanized operations such as pruning or the realization of the plantation. With the Super-Intensive System (SHD 2.0 SmartTree), it is possible to have a	Adoption of crops in line with local farming. Especially orchards produce a high remuneration per hectare (against high demands for labor) Supplementary income Introduction of areas dedicated to the cultivation of melliferous plant species for the breeding of honey bees (*Apis mellifera*) hosted in hives positioned under the photovoltaic	Adoption of crops in line with the local cultivations. Especially orchards produce a high remuneration per hectare (with high demands for labor). Supplementary income. Introduction of areas dedicated to the cultivation of melliferous plant species for the breeding of honey bees (*A. mellifera*), hosted in hives positioned under the photovoltaic

	Scenario 1	Scenario 2	Scenario 3	Scenario 4 (ANaV)
	Type of system			
Project characteristics	**Fixed mount photovoltaic**	**Intensive agri-voltaic (single-axis trackers + super intensive olive growing system)**	**Diversified agri-voltaic (single-axis trackers + diversified agriculture)**	**Agri-natural-voltaic (ANaV) (single-axis trackers + diversified agriculture and renaturalization)**
		greatly improved profitability, especially thanks to a significant reduction of laborers.	panels for ancillary honey production (Honey-Solar)	panels for ancillary honey production (Honey-Solar)
Integration with local agricultural production	None	Olive growing and the production of olive oil are typical of the project scope. This is a greatly developed agricultural sector in the agro-food district of Cerignola.	Existing local chain supplies for horticultural crops (agro-food district of Cerignola) and cereal crops (large-scale cultivation of durum wheat for pasta production in the area of Foggia)	Existing local chain supplies for horticultural crops (agro-food district of Cerignola) and cereal crops (large-scale cultivation of durum wheat for pasta production in the area of Foggia)
Employment	Total loss of jobs in the agricultural sector.	High loss of jobs in agriculture because of a high level of mechanization.	The adoption of crops chosen in line with local crops does not cause disturbances to the local market, including the labor market	The adoption of crops chosen in line with local crops does not cause disturbances to the local market, including the labor market

Table 2
Comparative table to understand the characteristics of different photovoltaic projects, in green the ANaV project.

cultivation in rows, that are well suited to the belt structure of the ANaV system. Similarly, the reduced height gain makes it possible to position some rows also in the area occasionally shaded by the tilting photovoltaic panels. Moreover, they do not require bulky machines that could damage the panels.

The agricultural cultivations envisaged by the ANaV project are economically effective in the agro-food district of Cerignola, with a supply chain of propagating material (plants), cold storage, processing rooms, and transport systems.

Crops have been positioned by dividing the full land into four macro-areas, based on the internal roads that actually allow the maneuvering of agricultural vehicles (internal roads are 10 m wide, so maneuvering is easy) [5].

Each macro-area is characterized by cultivation continuity (Artichoke, Asparagus, Cereals/Legumes), with the introduction of a belt of melliferous plants every 8 rows.

The belts cultivated with melliferous plants can also be crossed by vehicles since the species selected to give the ground strong stability and have good crushing resistance. For example, during the hand harvesting of the artichoke or the asparagus on these belts, trainers can be used for carrying the produce (**Figure** 7).

Figure 2.
Geographic position of Cerignola site of the ANaV project.

Considering the planned rotations, at the end of a profitable life of multiannual crops (artichoke and asparagus)—expected to be 5 years—the surface for artichoke cultivations will increase up to 60 hectares total (**Figure 8**).

The surface increase for artichokes rather than cereals/legumes also leads to a high increase in the profitability of the ANaV system thanks to the high remuneration of this horticultural production.

It is emphasized that the high innovation engendered by the ANaV system is due to the maximization of the cultivable surfaces in the installation area.

In fact, the fenced area of the ANaV system, corresponding to 141.28 hectares, is divided as follows:

- 124.28 ha for agricultural use;
- 10.80 ha are represented by a 1 m-wide belt under the photovoltaic modules that cannot be used for agricultural purposes since it is shaded by the modules for more than 6 hours a day;

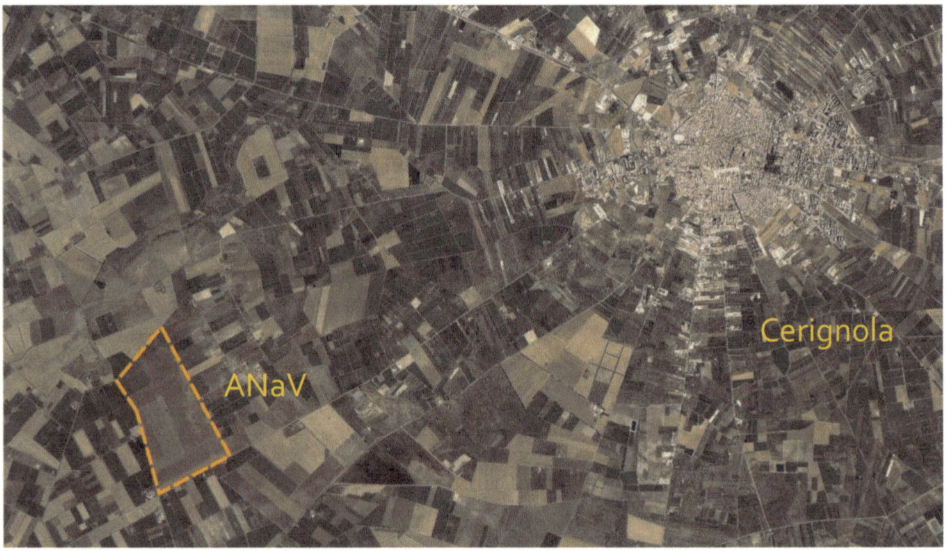

Figure 3.
Detail of the site of the ANaV project.

Figure 4.
Reconstructions of the ANaV environment.

- 5.39 ha are used by roads, electrical substations and by the water storage tank (for irrigation);

- 1.19 ha are occupied by a northern triangular area that will be used for agricultural activities and for the photovoltaic installation (a small office, a shelter for agricultural vehicles, storage of agricultural produce and lodgings for laborers).

Therefore, excluding the unproductive parts represented by roads, water storage tanks and different annexes, only 7.6% of the surface cannot be used directly for agricultural cultivations. However, it must be recalled that the hives are located on this area.

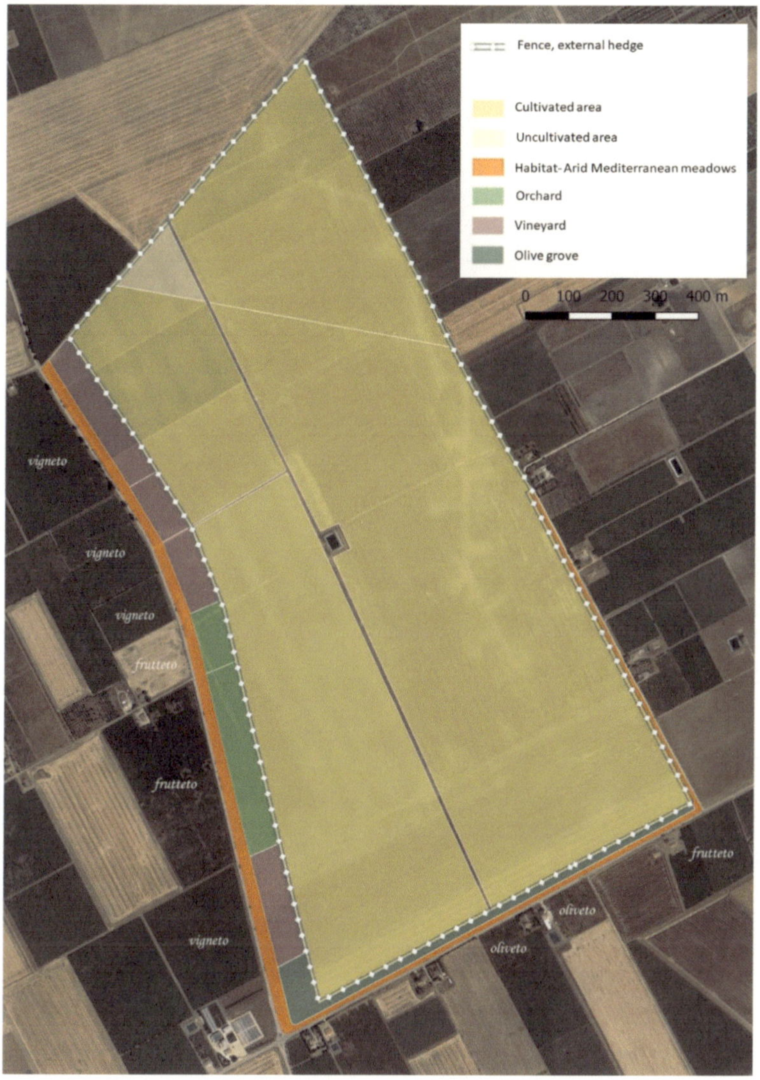

Figure 5.
Plan view of the ANaV project.

4. Conclusions

The ANaV project is made up of modules on "single-axis trackers" for a total capacity of 99.42MWp. The project has a usable agricultural area of 124.28 ha (89.32%) including the sub-panel belts and the green natural belts. Only 7.6% of the surface cannot be used directly for agricultural cultivations.

The ANaV project is an example of environmental "innovation" that shows it is possible to reconcile agricultural production, the increase of naturality, and photovoltaic power production in a sensitive geographic context. The Project allows a suitable realization of the initiative in the environmental context, safeguarding agricultural production and, simultaneously, positively influencing the botanic, vegetational, and wildlife contexts of the area.

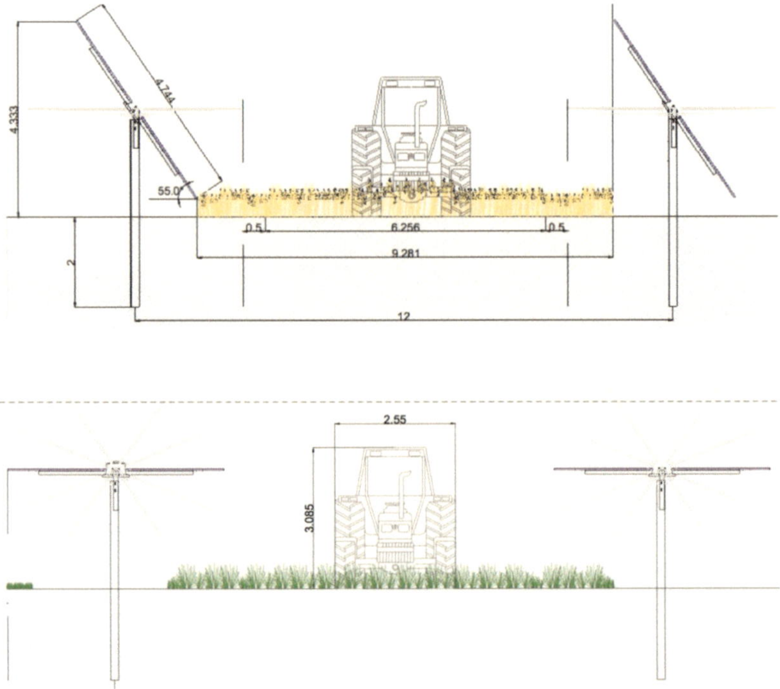

Figure 6.
Design dimensions for the ANaV system.

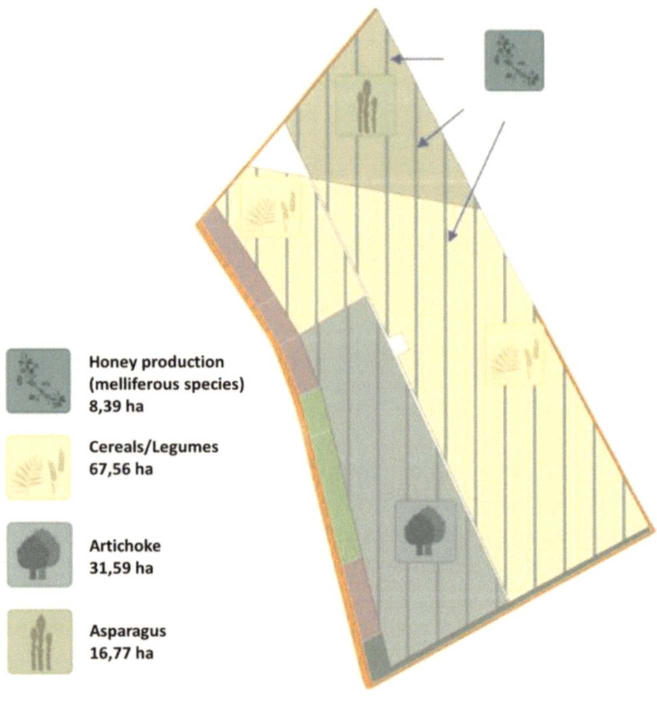

Figure 7.
Plan view of the crop layout in year 0.

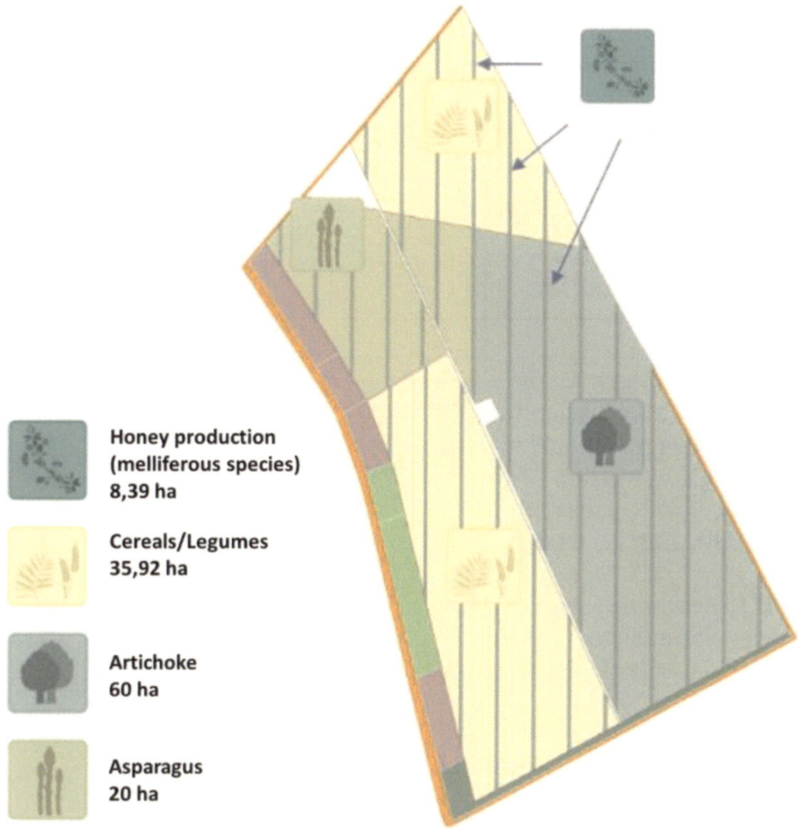

Figure 8.
Plan view of the crop layout in year 5.

Author details

Giovanni Campeol[1,2]*, Lorella Biasio[2], Silvia Foffano[2], Davide Scarpa[2] and Giulio Copparoni[3]

1 University Iuav of Venice, Italy

2 Studio ALIA, Treviso, Italy

3 Studio ALIA, Mogliano V.to, Italy

*Address all correspondence to: alia@aliavalutazioni.it

References

[1] Caffarelli A, Pignatelli A, de Simone G, Tsolakoglou K. Sistemi fotovoltaici. Progettazione Gestione Manutenzione impiantistica. Maggioli Editore; 2021

[2] Gallucci F. Colantoni A. e altri Linee Guida per l'applicazione agro-fotovoltaico in Italia; 2021. ISBN: 978-88-903361-4-0

[3] Toledo C, Scognamiglio A. Agrivoltaic systems design and assessment: A critical review, and a descriptive model towards a sustainable landscape vision (three-dimensional agrivoltaic patterns). Sustainability. 2021;**13**(12):6871

[4] Palchetti E. Relazione agronomica, impianto Agri-Naturalistico-Voltaico (ANaV). Foggia: Cerignola San Giovanni in fonte; 2021

[5] Campeol G. Strategic Environmental Assessment and Urban Planning: Methodological Reflections and Case Studies. Switzerland: Springer; 2020

Chapter 6

Valorization of Forest Waste for the Production of Bio-oils for Biofuel and Biodiesel

Hammadi el Farissi

Abstract

Biomass is a renewable energy source to generate heat and electricity through the enhancement of various organic materials. Cistus slow pyrolysis of seeds and shells was carried out in a fixed bed reactor to determine the effect of pyrolysis temperature, heating rate, and particle size on the performance of pyrolysis. Therefore, pyrolysis experiments were performed at different temperatures, ranging from 300 to 500°C, with heating rates varying from 10 to 70°C.min^{-1} for shells and 7 to 28°C.min^{-1} for seeds. The particle sizes of samples range from 0.3 to 3.5 mm for shells and 0.075 to 1.2 mm for seeds. The highest yield of liquid products (53.31% for shells; 52.24% for seeds) was obtained at a pyrolysis temperature of 450°C and a heating rate of 40°C. min^{-1} for shells and 21°C.min^{-1} for seeds. The functional groups and chemical compounds present in the bio-oil obtained under optimal conditions were identified by FTIR. The calorific value of the bio-oil was equal to 37.05 and 37.93 MJ.kg^{-1} for shells and seeds, respectively. The obtained results show that the bio-oil from the pyrolysis of Cistus shells and seeds could be used as a renewable fuel or a source of pharmaceutical and chemical raw material.

Keywords: bio-oil, bio-fuels, renewable energy, fixed bed pyrolysis, seeds of *Cistus Ladaniferus*, rockrose, green energy, viscosity

1. Introduction

The world's oil reserves are not eternal. Exploitation for fuel increases emissions of greenhouse gases that contribute to climate change. Renewable biomass is a promising alternative to petroleum-based products as a source of bio-energy and other bio-products [1], as the chemical value. The condensed Biogas is a type of liquid fuel made from biomass materials. As a kind of new cheap bio-energy, clean, green, bio-oil is considered an attractive option instead of conventional fuel in the aspect of reducing environmental pollution [2].

This study is motivated by a desire for development of natural resources to begin we take as cistus products. The study consisted of pyrolysis byproducts cistus ladanifer [3]. The cistus pyrolysis regenerates solid carbon-rich products (char) and condensable gaseous products (tar) and non-condensable (hydrocarbons). The process of pyrolysis is the newest part of renewable energy, has been set up and provides the benefits of a liquid product—bio-oil—which can be easily stored and transported and used as a fuel, vector energy and a source of chemicals. Bio-oils have been successfully tested in engines, turbines and boilers, and were up graded to high quality hydrocarbon fuels but currently unacceptably energy and financial cost [4].

The slow pyrolysis is often linked to coal production and fast pyrolysis has been linked to the production of bio-oil. The slow pyrolysis of biomass produces a high content of carbon [5]. Pyrolysis safflower seeds *(Carthamus tinctorius L.)* of particle size between 0.85 and 1.25 mm with a heating rate between 10 and 100°C.min^{-1} and the flow rate of nitrogen equals 100 cm^3.min^{-1}, the maximum yield was at 600°C for bio-oil and expensive parallel 53 and 17%. The effect of pyrolysis temperatures, the particle size and the rate of heating on the yield of the apricot kernel shells to pyrolysis temperatures range from 350 to 700°C with a heating rate ranges from 7 at 10°C.min^{-1}. For the lower heating rate of 10°C.min^{-1}, the carbonization yield is increased from 35.2 to 29.4%, as the final pyrolysis temperature was raised from 400 to 550°C. Same results were also observed by Gerc el [6], who studied the effects of different pyrolysis temperatures and heating rates on pyrolysis *(acanthium Onopordum L.)* When the temperature pyrolysis increased from 350 to 700°C the carbonization yield was increased from 38.3 to 24.1% with a rate of 7°C.min^{-1} heating. The decrease in the carbonization yield with increase in temperature could be due to either a larger primary decomposition of biomass at a temperature above or in the secondary decomposition of the tank [7–10]. The gas yield increased from 38 to 43%, with the same amount of increase in temperature. When the temperature increases [11]. *Yorgun et al* studied the fast pyrolysis of sunflower cakes in a tubular reactor. The effect of the final temperature, nitrogen flow rate and particle size on the performance of pyrolysis products were studied. The maximum yield of oil 45% by weight was obtained in the pyrolysis temperature of 550°C, with a flow rate of 1–300 cm^3.min^{-1} sweep gas and the particle size of 0425 to 0850 mm [12, 13].

After a series of studies of characterization of the seeds and shells of cistus by the different techniques of analysis, the determination of the percentages in carbon, oxygen, hydrogen, and nitrogen as well as the determination of the percentages in humidity, ash, fixed carbon, and volatile products, by the elementary analysis of the biomass, we give a clear idea on the pyrolysis of the used biomass. The knowledge of the above-mentioned data creates a good ground for the study of the different factors that influence the yield of bio-oil obtained by pyrolysis. In this chapter, we will study three main parameters that are: the effect of temperature, the effect of granulometry, and the effect of the heating speed on the yield of the two biomasses used.

2. Materials and methods

2.1 Raw materials

Biomass is the whole of the organic products, plants, and animals used for energy or agronomic purposes. The term biomass covers a very broad field: forestry waste, industrial waste, agricultural waste, the fermentable fraction of household waste and food industries, landfill biogas, or methanization products (sewage sludge, landfills, etc. ...). In this chapter, the different experimental campaigns were carried out using two types of biomass:

We took the initiative to study two types of biomass (**Figure 1**), one with little ash and the other with a lot more ash, in order to compare the different results and to be able to provide solutions for each type of biomass exploited. The main characteristics of the biomasses used are presented below.

2.2 Sample characterization

2.2.1 Determination of the moisture content

The moisture or water content designated by W of a sample is the ratio between the mass of water contained in the sample to its anhydrous mass, if we use its total mass this:

This ratio will be designated by W.

$$W = \frac{(m_0 - m_a)}{m_a} * 100 \tag{1}$$

Moisture is determined by subjecting a sample of known mass to oven drying (103°C) for 24 hours until the mass becomes constant. The humidity can be expressed as a percentage.

Relative to the anhydrous mass:

$$W_0 = \frac{(m_0 - m_a)}{m_a} * 100 \tag{2}$$

Or m_0: Total wet mass of the sample in (g); and m_a: Anhydrous mass in (g).

2.2.2 Determination of ash

The cistus seeds and these shells are impregnated samples were incinerated at 600° C, to a constant mass in a muffle furnace.

Figure 1.
(a) Plant with fruit; (b) seeds; (c) seed powder; and (d) plant with flower; (e) shells; and (f) shell powder.

The ash content, expressed as a percentage, is given by the equation:

$$\text{Ash}\ (\%) = \frac{(m_1 - m_{cr})}{(m_2 - m_{cr})} * 100 \tag{3}$$

Where, m_{cr}: Mass of the empty crucible (g); m_1: Mass of the crucible and the ashes (g); and m_2: Mass of the crucible and the biomass intake (g).

2.2.3 Determination of volatile matter and fixed carbon

The volatile matter represents the vapors of organic compounds and gases released by the biomass during pyrolysis, while the carbonaceous matter is the solid residue of carbon that remains after volatilization. Determination of volatile matter and fixed carbon for each sample was performed at 21°C.min^{-1} under an inert atmosphere for seeds and 40°C.min^{-1} for shells in a fixed bed pyrolysis reactor.

The volatile matter is determined by the formula:

$$\text{Mat.Vol}\ (\%) = \frac{(m_a - m_v)}{m_0} * 100 \tag{4}$$

Where: m_0: Initial mass of the sample; m_a: Mass of the dry sample (g); m_v: Mass of the devolatilized sample (g); m_C: Mass of ash (g).

The difference between the mass of the devolatilized sample (m_v) and that of the ash (m_C) represents the fixed carbon, designated by C. Fixed, whose mass percentage is given by the equation.

$$(\%)\ \text{C.Fixed} = \frac{(m_v - m_c)}{m_0} * 100 \tag{5}$$

The results of the various analyses are summarized in **Table 1**.

2.2.4 Elementary analyses

The values measured are comparable to the results obtained by the pyrolysis of olive stones [1, 2] and those obtained by the pyrolysis of castor oil [14–17]. The contents of sulfur and nitrogen obtained by the different samples are low compared to the other references. **Table 2** presents the elemental composition of pyrolysis by-products in the literature and the values obtained from different samples in this study.

According to the results obtained our biomass presents a significant percentage in C which is higher compared to other biomasses such as olive seed, and castor for

	Seeds (%)	Shells (%)
Moisture	13	7.3
Ash	9.6	2.94
Volatile matter	69.76	74.82
Fixed carbon	7.64	14.94

Table 1.
Characterization of different samples studied.

Literature	C	H	O	N	S
Olive Seeds *(Mehmet et al)*	47.36	6.04	45.52	0.96	0.12
Olive Kernel *(Elena Fernandez Ibañez)*	49.9	6.2	43.3	—	—
Castor seeds *(T.Hassan, R.Lakhmiri)*	59.25	7.15	29.94	3.20	0.46
Castor shells *(T.Hassan, R.Lakhmiri)*	49.8	5.3	43.9	0.9	0.1
Karanja Seeds [18]	53.04	7.32	35.53	3.94	0.18
Cistus Seeds	**68.7**	**4**	**26.4**	**0.74**	**0.16**
Cistus Shells	**69.03**	**3.75**	**26.49**	**0.68**	**0.05**

Table 2.
Elemental composition of same biomass compared with cistus seeds and shells.

example which does not exceed 59%, and for wood and coconut shells it is equal to 53.9 and 57.3% respectively [19]. On the other hand, the other elements like Oxygen, Nitrogen, Hydrogen, and Sulfur are lower than the biomasses quoted in **Table 2**. These values give the advantage to study the seeds and the shells of cistus for the production of bio-oils with the aim of transforming them into biofuels which are the vectors of actuality.

2.2.5 Analysis of the functions by FTIR

One introduces 2% of the sample with 98% of KBr then crushes the mixture to prepare pellets. We put the pellet in the support of infrared apparatus we obtain the spectra below.

Fourier transform infrared (FTIR) analysis of the raw material, seeds, and shells of cistus in **Figure 2** shows free -OH bonds at 3658.07 cm^{-1} and broadband of the bound

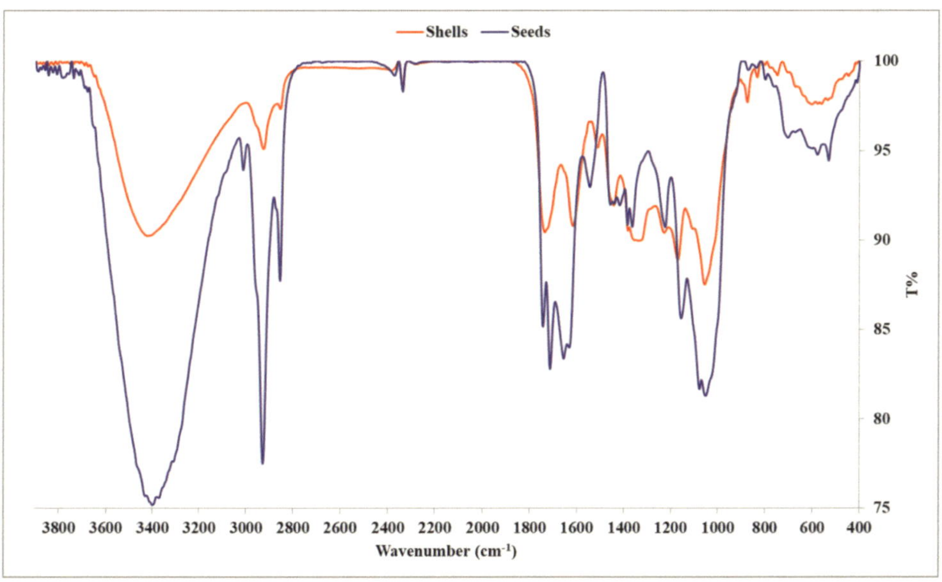

Figure 2.
FTIR spectrum of cistus shells and seeds.

-OH group, $=CH_2$ groups from 2866 to 3078 cm^{-1}, peaks of phosphene from 2280 to 2410 cm^{-1}, C=O bonds of the β-lactams with four centers from 1600 to 1720cm^{-1}, of aliphatic ketones 1705 to1725 cm^{-1}, C=C bonds of aromatics and phenols from 1550 to 1600 cm^{-1}, aromatic amines at 1515 cm^{-1}, C-O bonds of esters from 1210 to 1260 cm^{-1}, P-O-C bonds of phosphene at 1055 cm^{-1} and from 530 to 580 cm^{-1} of C-N bonds of nitriles, C-H bonds of mono and di-substituted aromatics from 650 to 900 cm^{-1} and finally the presence of cyclanes from 415 to 580 cm^{-1}. It can be concluded that the infrared spectroscopy confirms the important percentage of oxygen in the biomass by the presence of oxygenated groups like acids, Esters, Alcohols, Ketones, and Ethers.

2.2.6 Calorific values

The calorific values are calculated by a calorimetric bomb Leco501–053 Acetanilide. These values obtained are also compared to other biomasses. The results are given in **Table 3**.

According to the results of the **Table 3**, the seeds of cistus and their shells present a better calorific value of 25.12 $MJ.kg^{-1}$ and 23.29 $MJ.Kg^{-1}$ respectively. To our knowledge it is the highest in comparison with other biomasses cited in the literature and which are presented in **Table 3**, therefore, the biomass used is a very good source of bio-oil or biofuels of second generation for a sustainable environment and economy.

2.3 Experimental setup

2.3.1 Experimental procedure

The experiments were conducted to determine the influence of the pyrolysis temperature at a heating rate varying from 7 to 28°$C.min^{-1}$ for seeds and from 10 to 70°$C.min^{-1}$ for shells. The experiments were carried out in an apparatus designed with a cylindrical semi-batch reactor, in the shape of a vessel made of stainless steel, inserted vertically into an electrically heated oven (**Figure 3**). The temperature is controlled by a PID controller. The biomass sample (seeds or cistus shells) is introduced into the reactor during pyrolysis. The vapors generated from the reactor were condensed in a condenser cooled with chilled water. **Figure 3** represents the experimental setup.

After each experiment, the condensed liquid is collected in the cylindrical measuring device. After pyrolysis, the solid residue was collected and weighed. The sample

Material	Calorific Value $MJ.kg^{-1}$
Castor seeds [15]	24.47
Castor shells [16]	18.9
Black cumin seeds [20]	22.46
Karanja seeds [21]	22.38
Cistus seeds	**25.12**
Cistus shells	**23.29**

Table 3.
Calorific values of the raw material.

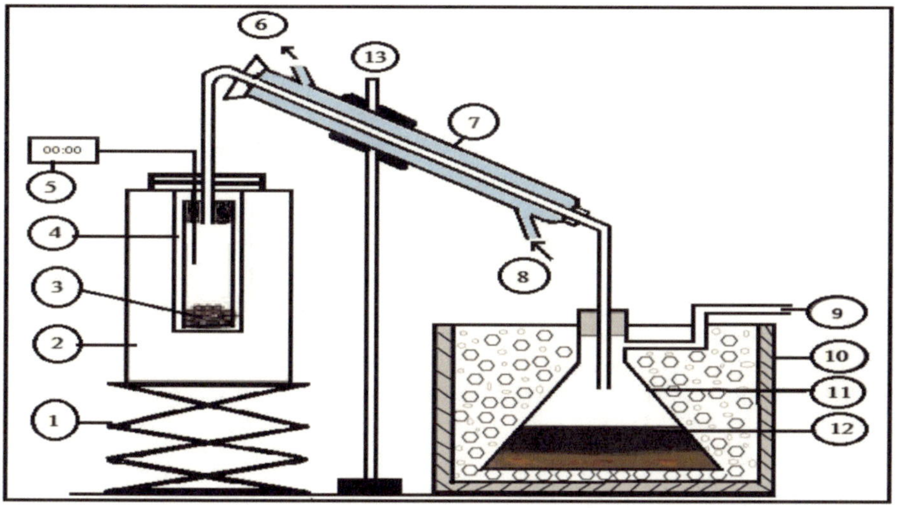

Figure 3.
Pyrolysis experimental set-up. Description: (1) elevator, (2) vertical tube farnace, (3) biomass sample, (4) pyolysis reactor, (5) temperature controller (PID), (6) water out, (7) condenser, (8) water in, (9) gas release, (10) ice bath, (11) condensate, (12) liquid, and (13) metal support.

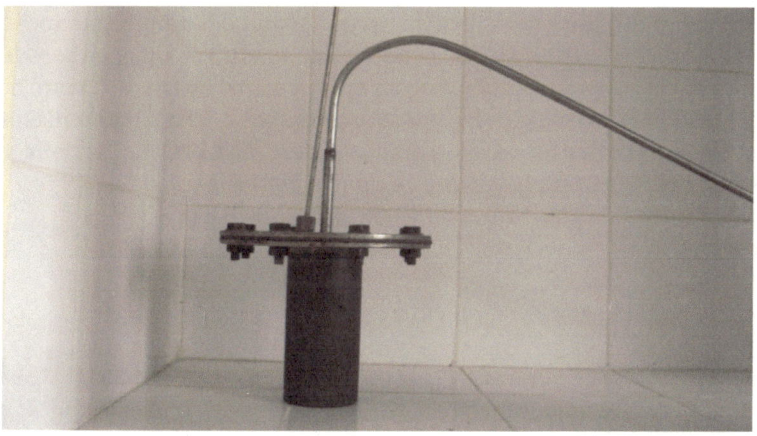

Figure 4.
Photography of reactor.

inlet of biomass and charcoal were solid measured by the electro weighing machine balance with an accuracy of ± 0.01 g.

2.3.2 Pyrolysis reactor

The reactor is a fixed bed (**Figure 4**). The detailed drawings are provided in part (2.3.1). It consists of a 310 AISI47 refractory stainless steel tube with an internal diameter of 60 mm and a total height of 150 mm. The head of the reactor is removable so that the bed can be introduced and then recovered at the end of the experiment. The whole reaction zone, between the diffuser and the gas outlet at the top of the reactor, has a total height of 150 mm. It includes the bubbling bed and the disengagement zone.

3. Results and discussion

To carry out this study we are obliged to fix the other parameters like particle size and heating rate as well as residence time. We introduce 15 g of the cistus seeds into the reactor which is fed with an electric current. The different results will be presented in the following sections. The yields of different pyrolysis products are calculated by the difference between the initial weight and the final weight which is the solid (char) that remains in the reactor. The liquid (Bio-Oil) which is taken in a graduated cylinder and the percentage of gas escaped to the atmosphere were calculated by the following relationship:

$$\%\text{Gas} = 100 - (\%\text{Solid} + \%\text{Liquid}) \tag{6}$$

In this step the particle size is fixed between 0.3 and 0.6 mm, to determine the effect of temperature. Using the results of the first experiments we have done we have found that the best yields in bio-oils are expected at a temperature equal to 450°C. At this stage we have already determined the main factor which is the temperature, and then it only remains the speed of heating that we vary from 7 to 28°C.min^{-1}. So, we take this conclusion in hand and we begin the study by the variation of the speed of heating from 7 to 28°C.min^{-1} each time, we calculate the yield of the obtained bio-oil. We deduced that for a heating speed lower than 21°C.min^{-1} we have yields of bio-oil less than 52.24% the same thing for speeds higher than 21°C.min^{-1}. Finally, we find that the optimal speed for this study is equal to 21°C.min^{-1}.

The diameter of the particles is fixed between 0.3 and 0.6 mm. 15 g of rockrose seeds are introduced into the fixed bed reactor and the temperature is changed between 300 and 500°C. The yields of the products of pyrolysis are calculated by the difference between the initial weight and the final weight that constitutes the solid (char) that remains in the reactor, the liquid (HP) that is taken in a graduated test tube, and the percentage of the gas that has escaped in the atmosphere were calculated by the relation of Eq. (6).

3.1 Cistus seeds

3.1.1 Effect of temperature on pyrolysis yield with d_p ($0.3 \le d_p \le 0.6$ mm)

Figure 5 represents the pyrolysis yields of cistus seeds with a heating rate of 21°C. min^{-1} starting from a temperature of 300°C. The yields of charcoal, pyrolysis oil and gas are quite close. When the temperature increases from 300 to 400°C we observe, on the one hand, an increase of liquid from 34.7 to 51.76%, and on the other hand a slight decrease of coal yield and gas yield down to 18.2%. In the range of 400 to 425°C, the liquid continues to increase along with a decrease in gas yield. At temperatures between 425 and 475°C a plateau of yield for the three (Bio-Oil, solid and gas) was noticed with a maximum yield of liquid at 450°C which is equal to 52.24%. From 475 to 500°C we observe a drop in the yield of solid and liquid in parallel and an increase in the yield of gas.

For the sizes that vary between 0.3; 0.4 and 0.5 mm at the same temperature we notice a small variation in the yield of pyrolysates which is equal to 0.2%. For this reason we have taken the particle size between 0.3 and 0.6 mm to complete the study.

3.1.2 Effect of heating rate on pyrolysis yields

Figure 6 shows the effect of heating rate on pyrolysate yields. The liquid and gas yields increase from 43.2 to 52.24% and from 27.60 to 17.51%, respectively when the

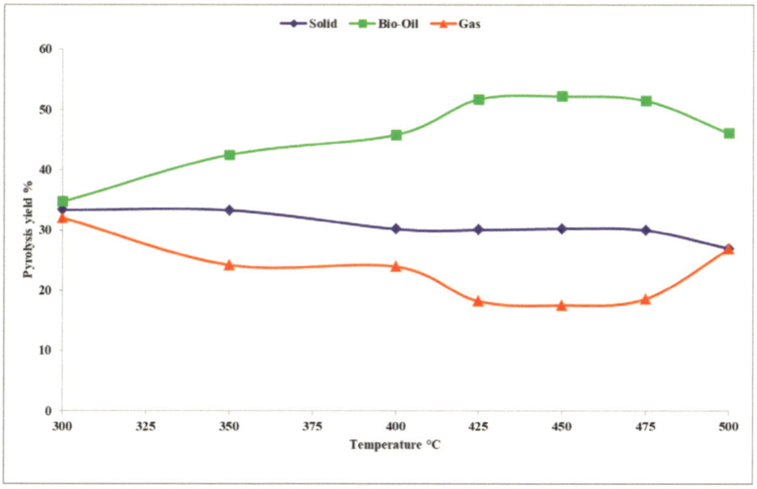

Figure 5.
Yield of pyrolysis products at various pyrolysis temperatures of seeds.

heating rate is increased from 7 to 21°C.min^{-1}. The increase in liquid yield with increasing heating rate may be due to higher heating rates breaking thermal barriers and mass transfer in the particles. The gas yield also increases with increasing heating rates due to the cracking of pyrolysis vapors at higher heating rates. The solid yield was increased slightly from 29.2 to 30.24 wt% when the heating rate was increased from 7 to 21°C.min^{-1}.

3.1.3 Effect of particle size on pyrolysis yields

The effect of particle size on the yields of pyrolysis products is shown in **Figure 7**, under the temperature of 450°C and the heating rate equal to 21°C.min^{-1}. Oil yield increased from 38 to 43.07% followed by a slight decrease in solid yield from 33.4 to

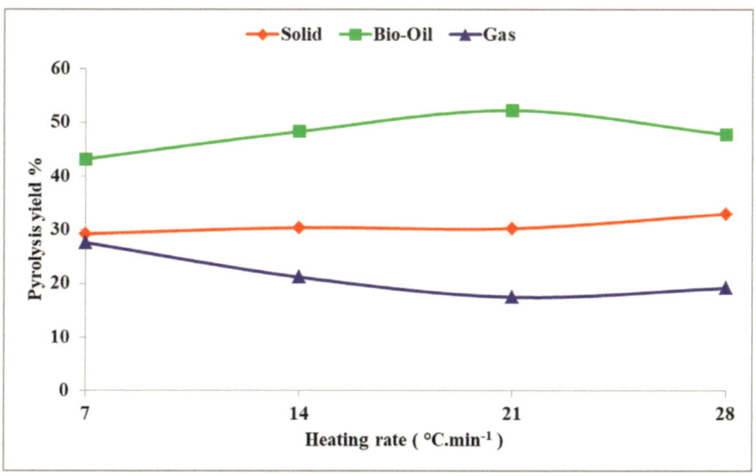

Figure 6.
Effect of heating rate on pyrolysis yields of seeds.

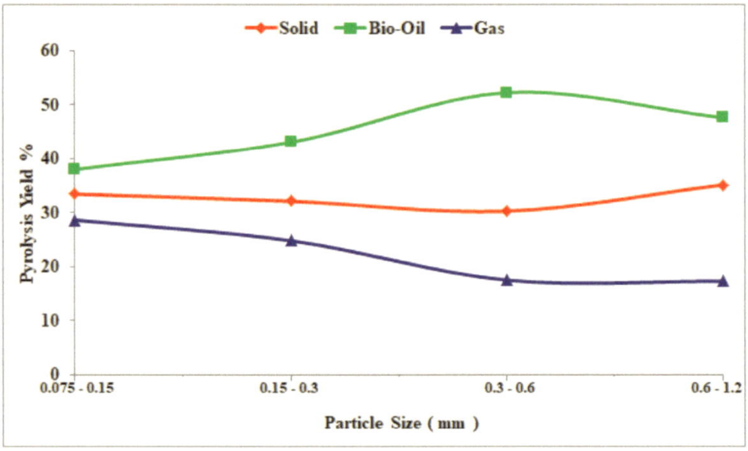

Figure 7.
Yield of pyrolysis products with different particle size seeds.

32.11 wt% and gas yield decreased from 28.6 to 24.82 wt% when the particle size increased from less than 0.075 to 0.15 mm and from 0.15 to 0.3 mm. And for particle sizes from 0.3 to 0.6 mm, the bio-oil yield follows a progressive increase until the maximum yield of 52.24%.

On the other hand the yield of gas and solid decreases from 24.82 to 17.51% and from 32.11 to 30.24% respectively. For particle sizes from 0.6 to 0.9 mm, the solid yield increases to 35.1%, and the gas and bio-oil yields decrease from 17.51 to 17.3% and from 52.24 to 47.6% respectively. For smaller particle sizes, the yield favors cracking hydrocarbons. The increases in solid yield with increasing particle size for the biomass sample could be due to a greater temperature gradient, within the particles. Thus at some point, the core temperature is lower than the surface temperature, this eventually gives rise to an increase in solid yield.

3.2 Cistus shells

3.2.1 Effect of temperature on the pyrolysis yield

Figure 8 shows the slow yields of pyrolysis products of cistus shells with the particle size of 2–3 mm at different temperatures from 300 to 500°C. The liquid yields and gas increased 38.53 to 44.6% by weight and 23.64 to 26.49%, respectively, while the solid yield decreased from 47.82 to 28.91% when the pyrolysis temperature is increased from 300 to 400°C.

In the temperature range of 400 to 450°C, it is observed a small decrease in solid and gas pass yield of 28.91 to 25.18% and from 26.49 to 21.51%, respectively, and the yield of bio-oil follow the increase maximum yield which was 53.31%. The low yield of liquid and low temperature gas is due to the incomplete decomposition of the shell. The decrease in the bio-oil yield and the increase in the gas yield of 47.11 and 28.6% respectively were observed at 500°C could be due to secondary cracking pyrolysis vapor and solid char. Similar results were observed in the study slow pyrolysis fixed bed of *Carpinus Betulus* residues *(U. Morali et al)*, the same trend was predicted by other researchers *(Kar Yakup and Ilknur Demiral)* different Pyrolysis of the liquid could be due to different biomass components.

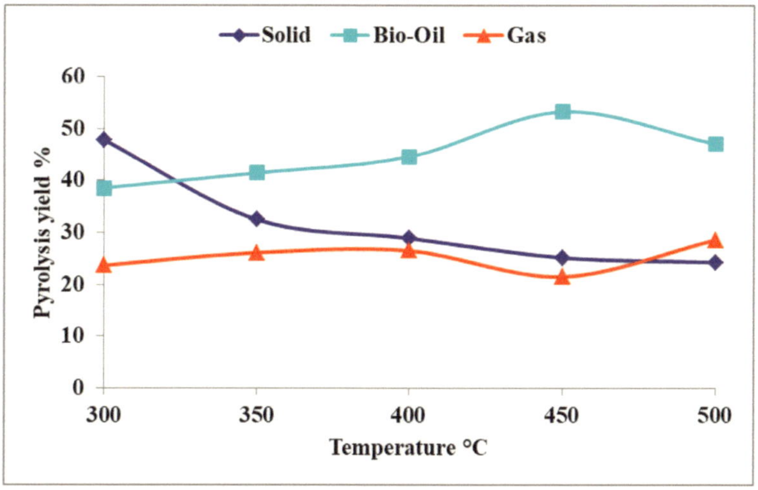

Figure 8.
Yield of pyrolysis products at various pyrolysis temperatures of shells.

3.2.2 Effect of particle size on pyrolysis yields

The effect of particle size on products yields was assessed by running pyrolysis experiments with a final temperature of 450°C and heating rate equal to 40°C.min^{-1}. Results are summarized in **Figure 9**. The lowest oil yield (35%) was obtained using the feedstock of tiniest particles (0.3–0.6 mm), which conversely afforded the maximal amount of gas products (38%) and a charcoal yield equal to 27%. When the particle size was increased to 1–2 mm, the oil yield was incremented to 48%, while both gas (27%) and char (25%) yields decreased.

The peak of oil production (yield = 53.31%), in conjunction with a further decline of gas (20%) and char (26.69%) yields, was achieved using 2–3 mm particles. Interestingly, the formation of charcoal (yield = 32.1%) reached a maximum when biggest

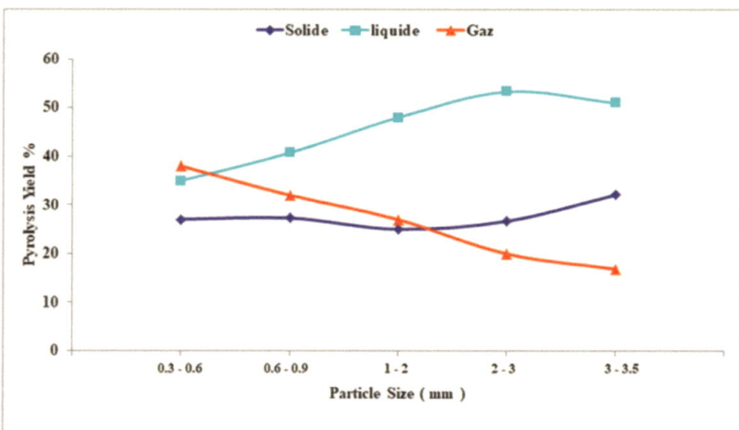

Figure 9.
Yield of pyrolysis products with different particle size shells.

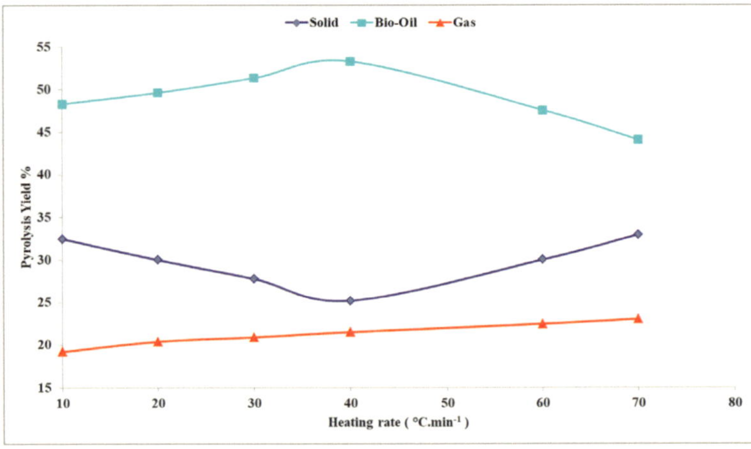

Figure 10.
Effect of heating rate on pyrolysis yields of shells.

particles (3–3.5 mm) were used, showing a significant effect of particle size on the performance of carbonization processes. On one hand, the use of smaller particles could promote the cracking of hydrocarbons, and the longer residence time of volatiles in the reactor would lead to the decrease of liquid yield. On the other hand, the increase of biomass particles size could produce a larger temperature gradient within the particles, so that at some point, the core temperature is lower than the surface, which might possibly lead to an increase in solid products yield.

3.2.3 Effect of heating rate on pyrolysis yields

Pyrolysis of cistus shells with particle size from 2 to 3 mm was next performed with a final temperature of 450°C and different heating rates. As shown in **Figure 10**, both oil and gas yields evenly grew upon increasing the heating rate from 10 to 40°C. min^{-1}, passing from 48.33 and 19.20%, respectively, at 10°C.min^{-1} to 53.31 and 21.5%, respectively, at 40°C.min^{-1}. The char yield dropped from 32.47 to 25.19%. This change can be ascribed to the shorter residence time and reduced incidence of cracking for pyrolysis vapors, which also account for the increased yield of tar. At 40°C. min^{-1}, the optimal heating rate for the production of oil (yield = 53.31%), the yields of charcoal and gas underwent a further slight reduction. Finally, for heating rates greater than 40°C.min^{-1}, decreased oil yield against increased solid and gas yields due to the fast pyrolysis of cistus seeds were observed.

3.3 Bio-oil

The obtained bio-oil is characterized by FTIR spectroscopy to determine the different functional groups existing. **Figures 11** and **12** shows the infrared spectrum of the bio-oil of cistus seeds and shells. **Tables 4** and **5** represents the results of elemental analysis of the bio-oil and the calorific values.

The analysis of the bio-oil of cistus seeds by Fourier Transform Infrared (FTIR) (**Figure 11**) shows the C-H bonds of cyclobutane in symmetrical and antisymmetrical vibration between (2800–2900 cm^{-1}), the aliphatic ketones (1705–1725 cm^{-1}), the C=C bonds of aromatics and phenols between (1550–1600 cm^{-1}), the C-O bonds of

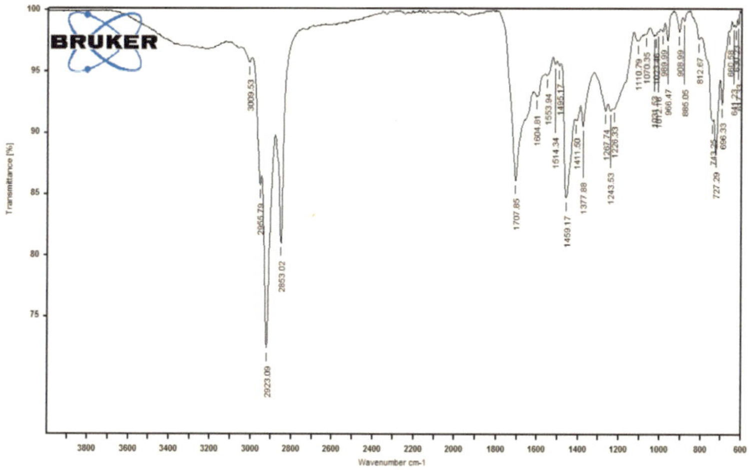

Figure 11.
FTIR spectrum of bio-oil (Cistus seeds).

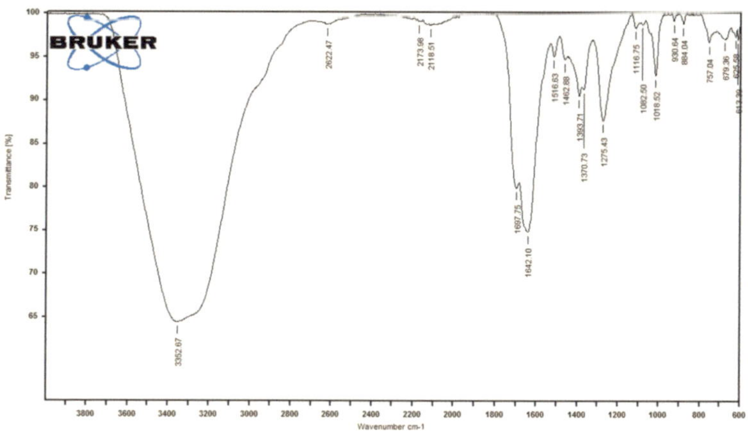

Figure 12.
FTIR spectrum of bio-oil (Cistus shells).

esters between (1210–1260 cm^{-1}), C-OH bonds of primary and secondary alcohols between (1050–1080 cm^{-1}), and between (1110–1220 cm^{-1}), C-N bonds of aromatic amines between (1020–1220 cm^{-1}), C-H bonds of mono and disubstituted aromatics between (650–900 cm^{-1}) and finally the presence of (Z) and (E) isomers of alkenes between (650–750 cm^{-1}) and (950–1010 cm^{-1}).

Bio-oil analysis of cistus shells by oven transformation (FTIR) (**Figure 12**), shows -OH groups of phenols and acids between at 3352.67 cm^{-1} and 2622.47 cm^{-1} respectively, the C=O of acids at 2173.98 cm^{-1} and aromatic ketones at 1697.75 cm^{-1}, the $C \equiv C$ of alkynes at 2118.51 cm^{-1}, the C=C of alkenes at 1642, 10 cm^{-1} and, the C-N of amines at 1516.63 cm^{-1}, the C-H of aldehydes at 1370.73 cm^{-1}, the C-N of aromatic amides at 1275.43 cm^{-1}, the C-O of esters and ethers at 1082.50 cm^{-1}, the nitrile groups -NO_2 at 930.64 cm^{-1} and finally the C-H bonds of polycyclic and substituted aromatic groups between 679 and 757.04 cm^{-1}.

Element	Ultimate analysis (w/w %)						
	Cistus seeds [22]	Castor [15]	Sesame [23]	Mustard	Neem	Jatropha	Rapeseed
C	**71.5**	70.5	45.19	47.47	42.52	59.17	45.92
H	**11.85**	10.35	7.55	5.73	4.52	6.52	6.21
N	**0.63**	5.1	7.26	6.16	1.87	0.38	6.90
S	**NI**	NI	0.72	1.74	1.30	NI	0.88
O	**16.02**	14.05	39.27	38.91	49.79	33.93	40.09
H/C	**1.99**	1.75	0.17	1.45	1.28	NI	1.62
O/C	**0.17**	NI	NI	NI	NI	NI	NI
HV(MJ/kg)	**37.93**	36.5	19.78	20.5	18.20	13.55	19.84
Formula	**$CH_{1.99}O_{0.17}N_{0.01}$**						

Table 4.
Elemental analysis of bio-oil from seeds.

Element	Ultimate analysis (w/w %)				
	Cistus Shells	Castor [16]	Apricol Kernel [11]	Hornbeam (Charme) [24]	Walnut (Noix) [25]
C	**71.7**	74.5	64.45	66.42	59.89
H	**14.87**	10.76	8.24	6.93	7.33
N	**0.54**	2.56	0.81	1.54	0.5
S	**NI**	NI	NI	NI	0.02
O	**12.89**	11.78	26.50	25.11	32.26
H/C	**2.49**	1.76	1.53	1.24	1.74
O/C	**0.21**	0.12	NI	NI	NI
Calorific Value (MJ/kg)	**37.05**	35.01	27.19	30	25.01
Formula	**$CH_{2.49}O_{0.21}N_{0.01}$**	NI			

Table 5.
Elemental analysis of bio-oil from shells.

The results of **Table 5** shows that our bio-oil extracted by fixed bed pyrolysis contains more carbon compared to Apricol, hornbeam and walnut shells, but the percentage of oxygen in the bio-oil is smaller than Apricol, hornbeam and walnut bio-oils. In addition, we observe a total absence of sulfur in our bio-oil. We do not forget that the energetic value of the bio-oil is the highest compared to the bio-oils quoted in the literature and close to that of the oil which varies from 40 to 44 $MJ.kg^{-1}$ (**Table 6**).

The second method for the measurement of the kinematic viscosity is obtained by measuring the time of flow of a given volume of liquid under the effect of gravity (the dynamic viscosity (η) in ($g.cm^{-1}.s^{-1}$ or mPa.s) of a fluid is obtained by multiplying its kinematic viscosity (γ) in ($cm^2.s^{-1}$ or stokes (cSt)) by its density (ρ)). Experimentally

Samples	Density	Speed limit	A	B	Dyn. Viscosity	Kin. Viscosity
	mg cm^{-3}	cm.s^{-1}	cm.s^{-2}	s^{-1}	η (mPa.s)	η (mm^2.s^{-1})
Bio-oil (seeds)	897.6	1.0182	6.5487	6.4317	3.8693	4.3107
Bio-oil (shells)	881.3	1.0521	6.6078	6.2806	3.7784	4.2871
Commercial Diesel 1	829.9	1.2109	6.7947	5.6113	3.3758	4.0677
Commercial Diesel 2	829.4	1.2268	6.7965	5.5400	3.3329	4.0185
Commercial Diesel 3	830.3	1.2475	6.7931	5.4454	3.2760	3.9454

Table 6.
Properties physicochemical experimental obtained by the mechanic study.

Samples	Time	Distance	Dyn Viscosity	Kin Viscosity	Volume
	t (s)	d (cm)	γ (mPa.s)	η (mm^2 s^{-1})	V (cm^3)
Bio-oil (Seeds)	0.116	20	3.869	4.310	10
Bio-oil (Shells)	0.118	20	3.734	4.237	10
Commercial Diesel 1	0.126	20	3.293	3.968	10
Commercial Diesel 2	0.125	20	3.318	4000	10
Commercial Diesel 3	0.128	20	3.243	3.906	10

Table 7.
Properties physicochemical experimental obtained by the volume study.

not having a viscometer the measurement of the time of flow of a volume V = 10 cm^3 of liquid in a graduated burette of length 20 cm, the temperature of 20 °C the results are grouped in **Table 7** gave: $\eta = \gamma/\rho$ with ρ in g.cm^{-3} [26].

We can deduce from the unit of kinematic viscosity the relationship that links the volume of liquid, the time of decantation and the length of the burette $\gamma = V/L.t$ These results allow to highlight the small difference in viscosity between the bio-oil of seeds and shells of cistus and biodiesel. So we can improve the density of our bio-oil by

Physical properties	Cistus		Commercial		
	Seeds	Shells	Diesel 1	Diesel 2	Diesel 3
Appearance	Typically a dark brown free flowing liquid		Yellowish		
Odor	A distinctive smoky smell		Aromatic		
Calorific value (MJ Kg^{-1})	37.93	37.05	41.50	43.20	42.35
Density (mg cm^{-3})	897.6	881.3	829.9	829.4	830.3
Dyn Viscosity (mPa.s)	3.768	3.716	3.306	3.255	3.254
Kin Viscosity (mm^2 s^{-1})	4.198	4.217	3.986	3.922	3.919
pH	3.8	4.02			
Miscibility	Methanol, Ethanol, Toluene, diesel, and Petrol				

Table 8.
Fuel properties of cistus seeds and shells pyrolysis oil.

adding a percentage of ethanol to increase the calorific value and thus bringing its density closer to the density of commercial fuels. To improve the bio-oil we can refine it by the reactions of trans-esterification to mimic the acids and the methyl esters of the vegetable oils as well as the ethyl ethers.

To confirm the experimental results found in this study concerning viscosity and density, measurements results are obtained at 20°C, were made using an apparatus (Anton paar DMATM4500M) with a (Software Version 2.93.9364.129) (**Table 8**).

4. Conclusion

A parametric study focused mainly on the impact of temperature, size and heating rate on the yield of pyrolysates. The ex-beech liquids show high solid residue and gas contents, good homogeneity and yields up to 52.24% for seeds and 53.34% for shells at 450°C. Chemical analyses were also carried out to characterize the pyrolysis products obtained in order to determine the oxygen, carbon and hydrogen contents in the solid residues and in the liquids.

So far, the percentage of C, O, N and H of the pyrolysis oils as a whole is high compared to other biomasses located in the literature and also present important and very high calorific powers in comparison with wood and with other biomass as castor, black cumin, karanja, apricol, walnut and hornbeam The main innovative character of this study lies in the adopted approach which consists in valorizing the bio-oils of pyrolysis, in particular in the production of biofuels in a first time and in a second time the synthesis of chemical products with the aim of use in cosmetic, pharmaceutical and food. The valorization of the solid (charcoal) as a bioadsorbent which will be detailed in another chapter.

Acknowledgements

This work is the result of research on the valorization of bioresources, it is a chapter of my doctoral thesis.

Conflict of interest

The author declares no conflict of interest.

Institutional Review Board Statement

Not applicable.

Informed Consent Statement

Not applicable.

Data Availability Statement

Not applicable.

Author details

Hammadi el Farissi
Faculty of Sciences, Department of Chemistry, Laboratory of Environment and Applied Chemistry (LCAE), Physical Chemistry of the Natural Resources and Processes, Mohamed First University, Oujda, Morocco

*Address all correspondence to: hammadielfarissi04@gmail.com

References

[1] Ksel Y, Yanik J, Kommayer C, Sağlam M. Fast pyrolysis of agricultural wastes. Characterization of pyrolysis products. Fuel Processing Technology. 2007;**88**:942-947

[2] Riehle P, Vollmer M, Rohn S. Phenolic compounds in Cistus incanus herbal infusions—Antioxidant capacity and thermal stability during the brewing process. Food Research International. 2013;**53**(2):891-899. DOI: 10.1016/j.foodres.2012.09.020

[3] Dentinho MTPTP, Belo ATT, Bessa RJBJB. Digestion, ruminal fermentation and microbial nitrogen supply in sheep fed soybean meal treated with Cistus ladanifer L. tannins. Small Ruminant Research. 2014;**119**(1–3):57-64. DOI: 10.1016/j.smallrumres.2014.02.012

[4] Pattiya A. Bio-oil production via fast pyrolysis of biomass residues from cassava plants in a fluidised-bed reactor. Bioresource Technology. 2011;**102**(2): 1959-1967. DOI: 10.1016/j.biortech.2010.08.117

[5] Onay O. Influence of pyrolysis temperature and heating rate on the production of bio-oil and char from safflower seed by pyrolysis, using a well-swept fixed-bed reactor. Fuel Process Technology. 2007;**88**(5):523-531

[6] Gercel HF. Bio-oil production from *Onopordum acanthium* L. by slow pyrolysis. Journal of Analytical and Applied Pyrolysis. 2011;**92**:233-238

[7] Demiral I, Eryazıcı A, Sensöz S. Bio-oil production from pyrolysis of corncob (*Zea mays* L.). Biomass Bioenergy. 2012; **36**:43-49

[8] Horne PTWPA. Influence of temperature on the products from the flash pyrolysis of biomass. Fuel. 1996;**75**: 1051-1059

[9] Jerónimo E et al. Effect of dietary grape seed extract and Cistus ladanifer L. in combination with vegetable oil supplementation on lamb meat quality. Meat Science. 2012;**92**(4):841-847. DOI: 10.1016/j.meatsci.2012.07.011

[10] Oladoja NA, Aboluwoye CO, Oladimeji YB, Ashogbon AO, Otemuyiwa IO. Studies on castor seed shell as a sorbent in basic dye contaminated wastewater remediation. Desalination. 2008;**227**(1–3):190-203. DOI: 10.1016/j.desal.2007.06.025

[11] Pyrolysis A, Engineering C. Pyrolysis of apricot kernel shell in a fixed-bed reactor : Characterization of bio-oil and char. Journal of Applied Analysis. 2014; **107**:17-24. DOI: 10.1016/j.jaap.2014.01.019

[12] Couic-Marinier F, Lobstein A. Les huiles essentielles gagnent du terrain à l'officine. Actualités Pharmaceutiques. 2013;**52**(525):18-21. DOI: 10.1016/j.actpha.2013.02.005

[13] Sensoz S, Angin D, Yorgun S. Influence of particle size on the pyrolysis of rapeseed (*Brassica napus* L.) :fuel properties of bio-oil. Journal of Biomass and Bioenergy. 2000;**19**:271-279

[14] Ibañez MEF, Etude de la carbonisation et l'activation de précurseurs végétaux durs et mous, 2002

[15] Mohammed TH, Lakhmiri R, Azmani A. Bio-oil from Pyrolysis of Castor seeds. International Journal of Basic Applied Science. 2014;**14**(04):14-17

[16] Mohammed TH, Lakhmiri R, Azmani A, Hassan II. Bio-oil from

Pyrolysis of Castor Shell. International Journal of Basic Applied Science. 2014; **14**(06):1-5

[17] Singh RK, Shadangi KP. Liquid fuel from castor seeds by pyrolysis. Fuel. 2011;**90**(7):2538-2544. DOI: 10.1016/j.fuel.2011.03.015

[18] Fagbemi L, Khezami L, Capart R. Pyrolysis products from different biomasses. Applied Energy. 2001;**69**(4): 293-306. DOI: 10.1016/S0306-2619(01)00013-7

[19] Shadangi KP, Mohanty K. Thermal and catalytic pyrolysis of Karanja seed to produce liquid fuel. Fuel. 2014;**115**(July): 434-442. DOI: 10.1016/j.fuel.2013.07.053

[20] Şen N, Kar Y. Pyrolysis of black cumin seed cake in a fixed-bed reactor. Biomass and Bioenergy. 2011;**35**(10): 4297-4304. DOI: 10.1016/j.biombioe.2011.07.019

[21] Nayan NK, Kumar S, Singh RK. Characterization of the liquid product obtained by pyrolysis of karanja seed. Bioresource Technology. 2012;**124**: 186-189. DOI: 10.1016/j.biortech.2012.08.004

[22] El Farissi H, Lakhmiri R, El Fargani H, Albourine A, Safi M. Valorisation of a forest waste (Cistus seeds) for the production of bio-oils. Journal of Materials Environmental Science. 2017;**8**(2):628-635

[23] Volli V, Singh RK. Production of bio-oil from de-oiled cakes by thermal pyrolysis. Fuel. 2012;**96**:579-585. DOI: 10.1016/j.fuel.2012.01.016

[24] Ugur Moralı SS. Pyrolysis of hornbeam shell (*Carpinus betulus* L.) in a fixed bed reactor: Characterization of bio-oil and bio-char. Fuel. 2015;**150**: 672-678. DOI: 10.1016/j.fuel.2015.02.095

[25] Kar Y. Co-pyrolysis of walnut shell and tar sand in a fixed-bed reactor. Bioresource Technology. 2011;**102**(20): 9800-9805. DOI: 10.1016/j.biortech.2011.08.022

[26] El Farissi H et al. Characterization of bio-oils synthesized by pyrolysis from cistus ladaniferus transformed into biofuel and biodiesel used as an alternative green energy. In: Second International Conference on Advanced Materials Behaviour. Vol. 2417. ICAMBC; 2021. p.060003. DOI: 10.1063/5.0072580

Section 3

Green Healthcare and Computing Systems

Chapter 7

Going Green in Ophthalmic Practice

Prasanna Venkatesh Ramesh, Shruthy Vaishali Ramesh, Prajnya Ray, Aji Kunnath Devadas, Akshay Surendran, Tensingh Joshua, Meena Kumari Ramesh and Ramesh Rajasekaran

Abstract

The healthcare sector has had a relatively late implementation of environmental thinking in its setup. "Green electronic technologies and go green healthcare system" has recently evolved to address the biodegradability and biocompatibility issues faced by inorganic electronics and non-biodegradable materials. Green healthcare has the capability to promote global health both directly and indirectly. With the rise in environmental degradation, utilizing innovative 'Go Green' strategies in ophthalmology is of utmost importance. It has been structured to imbibe environmentalism into healthcare. In this chapter, we have emphasized a few simple patient point of care (POC) innovations in the field of ophthalmology that could transform the future of disease management toward a much more sustainable model by reducing resource and energy consumption. We have discussed how we innovated the novel coronicle (corona + cubicle) during the COVID-19 era, which housed all the essential ophthalmic gadgets and was interconnected using the local area network (LAN) for data access in patient care. Turning to patient counseling, we have highlighted how our innovative and cost-effective 3D augmented reality, and 4D holographic diagnostics and counseling platforms are effective in replacing the conventional paper-based system.

Keywords: green electronic technologies, patient point of care, 3D augmented reality, 4D holographic diagnostics, counseling platforms, go green, ophthalmic cubicle

1. Introduction

Environmental changes are considered by many as the major long-term threat to global health in the 21st century [1]. To keep an eye on that, worldwide communities, governmental agencies or international research programs like Green Program 2030 have made massive, concerted efforts to launch new visions in the economy, society, and healthcare sector such as green building, green cities and other go green initiatives, but, global environmental issues and its potentially catastrophic effects are accelerating faster than anticipated [2–6]. Healthcare services are more substantial contributors to climatic changes as it generates the most diverse both non-biodegradable and biodegradable biohazard waste materials in large quantities; compared to any other commercial sector. At least 15% of that are highly hazardous and not managed in an environmentally safe manner [7, 8]. In addition, the percentage of plastic in medical waste may be as high as about 20–30% [9]. Moreover, since the first outbreak of coronavirus disease (COVID-19) early in 2020, eight million tons of pandemic-associated waste (plastic) have been generated globally, contributing to environmental degradation and climatic change [10–12]. It is a known fact that healthcare services are significant contributors to total national greenhouse gas emissions, such as 10% in the United States of America (USA), 7% in Australia, 5% in Canada and Japan, 4% in the United Kingdom (UK) and 1.5% in India [3, 13–17].

Without question, modern eye care practice, as a high-volume service, generates an enormous amount of non-biodegradable trash regularly, and it all starts with any organization's routine surgical and outpatient department (OPD) procedures, whether in private practice or academia or somewhere in between. For instance, ophthalmology is a high-volume speciality, accounting for 8·1% of hospital outpatient visits nationally in 2018–2019 in the UK [18]. Consequently, all eye healthcare centres in the world face numerous challenges, including inaccessibility of good services, rising costs, and an increase in environment-related pollution.

The adverse effects of ophthalmic healthcare delivery on the environment will probably increase daily. There are many innovations implemented to solve this problem, and demand is rising as the world population grows and ages. In this chapter, we have discussed one such emerging green concept; by embracing this concept, eye hospitals mainly benefit in terms of energy-saving, which can also lead to monetary savings. The green color is indicative of the effort taken to create an eco-friendly healthcare system. Going green involves waste reduction, and energy and resource conservation in the modern healthcare system, which requires expensive, energy-intensive processes in the use of water, lighting, heating, cooling, ventilation, and waste disposal [19]. As per estimates, 15 to 30% of energy and around 30% of water could be saved in this process. Usage of general lighting, compact fluorescent lamps (CFLs) and light-emitting diode (LED) lamps can reduce energy consumption by 15–30%, 30% and 45–50%, respectively [12]. The Indian Green Building Council (IGBC) and Green Rating for Integrated Habitat Assessment (GRIHA) are the green building rating systems that help review the requirements and aid in setting up goals for green projects by targeting elements of sustainability [12].

The medical community has already implemented green practices in various surgical units, such as in cataract surgery and surgical waste disposal. Still, the functioning of green teams should be beyond the hospital setting when involving long-term care or outpatient clinics [20]. In this chapter, we have discussed simple patient point of care (POC) innovations for diagnosing eye diseases and counseling platforms toward a much more environment-friendly culture in the OPD.

2. Green outpatient department and clinics

A healthy work environment is essential in healthcare not only to care for its own but to model for society at large the value of a healthy environment maintained according to principles of sustainability [21]. The essential principles of green clinics are to create a workspace that is safe and effective for both the patient and health workers and provides a sustainable model for long-term global health.

2.1 Coronicle (Corona +cubicle)

With the emergence of the COVID-19 pandemic, many drastic changes were made in the ocular healthcare system to provide quality healthcare without compromising safety and functionality, while also considering greener strategies [22, 23]. One such innovative practice pattern was the ophthalmic coronicle (cubicle) which was constructed to ensure safety for both healthcare workers and OPD patients in a high-volume urban private set-up [24].

2.1.1 Establishment of the Coronicle

During the COVID-19 pandemic, the pre-COVID waiting hall of approximately 450 square feet was converted into a state-of-the-art investigation coronicle of 256 square feet for comprehensive ocular examination (**Figure 1**). Dimensions of the coronicle were 16 feet in length, 16 feet in breadth, and 8 feet in height; since only a minimal area was utilized for patient examination, the usage of electricity was reduced greatly, contributing to environmental sustainability. Acrylic sheets, aluminum beading, fevicol, araldite paste, and a jigsaw cutting blade were used to build the coronicle in a do-it-yourself (DIY) template. A detailed cost analysis of the materials and the total cost of setting up the coronicle are provided in (**Table 1**) [24]. There are two slit lamps, two electronic medical records (EMR) computer systems, and one auto refractometer, lensometer, confocal fundus capture device, spectral-domain optical coherence tomography (OCT), optical biometer, manual keratometry, corneal topography, and non-contact tonometer (NCT) encompassed inside the coronicle (**Figure 2**), which were interconnected via local area network (LAN). The coronicle has proved to be effective over a period of two years (June 2020–July 2022) with the majority of the patients only needing the services available within the coronicle ($p = 0.00$). **Table 2** provides the details of the number of patients requiring the services available within and outside the cubicle during this time period.

Figure 1.
Image showing the coronicle (corona + cubicle) in the pre-COVID waiting hall.

Materials	Dimensions	Price/unit (GBP)	Quantity	Amount (GBP)
Acrylic Sheet (8 mm thickness)	16*8 feet	4.61/square feet	4	2358.29
Aluminum Beading (4*4-inch thickness)	16 feet length		4	146.56
Aluminum Beading (2*2-inch thickness)	8 feet length		20	146.56
Screws, nuts, and bolts				18.84
Adhesives (fevicol + araldite)				26.17
Lock for sliders and door		8.37	9	75.37
Air Conditioner	2 tons	471.07	1	471.07
Labour charges		0.84/square feet		214.39
Cutting charges				39.78
Total				**3497.03**

Table 1.
Materials used for assembly the cubicle and their cost dynamics.

2.1.2. Go green with Coronicle

As all the ophthalmic instruments need maintenance via good air conditioners for their longer sustainability, keeping each instrument in a different room ends up with a great amount of electricity consumption as well as a higher number of manpower, which is a major contributor to the carbon footprint. Air conditioners release potent greenhouse gases into the air causing insulation of our planet; it is a major contributor to global warming. Before the implementation of the coronicle, the total electricity consumption of the hospital was approximately 22,000 kilowatts per month, which has effectively reduced after the implementation of the coronicle to approximately 13,000 kilowatts per month. So, instilling all the gadgets in a single coronicle can effectively reduce the usage of the number of air conditioners as well as the number of tube lights, which in turn can reduce the consumption of electricity and can lead to a climate-friendly and energy-efficient healthcare facility. This way eye care hospitals can take initiative to address global environmental health for the future and present-day generations.

2.2 Local area network (LAN)

2.2.1 Features of LAN

Similarly, during the pandemic, along with the coronicle, we were able to optimally utilize LAN, which interconnects all the ophthalmic gadgets to the personal computer (PC) (**Figure 3**). As a result, it amplifies the functionality of holistic eye examination using a single internet connection [25]. In cases with multi-modal imaging and testing, LAN plays a vital role by not only connecting the devices but also providing in-depth data without compromising on quality.

2.2.2 Setting up the LAN framework

The LAN networking is implemented by a sixteen-port switch (**Figure 4a**) present inside the coronicle which is connected to the CISCO (Commercial and Industry

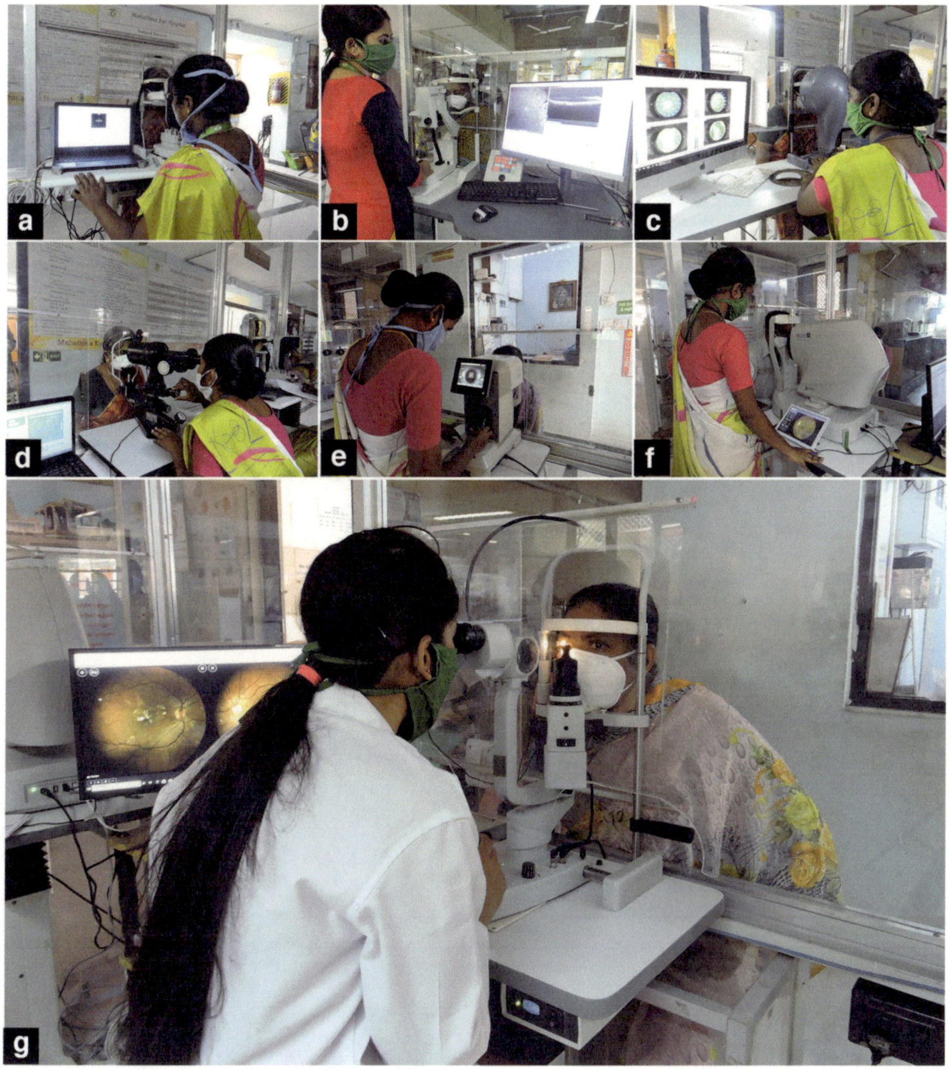

Figure 2.
All the essential ophthalmological instruments required for an eclectic setup inside the coronicle (a) optical biometry (b) optical coherence tomography, (c) corneal topography, (d) manual keratometry, (e) non-contact tonometer, (f) confocal fundus photography, and (g) slit lamp biomicroscopy.

Security Corporation) switch (**Figure 4b**) inside the server room placed in a 6 U rack. CISCO allows the connected devices to share information and communicate with each other on the same network inside the building for high-security purposes. LAN also link wireless access points, printers, xerox machine, scanners and servers on the same network for extra facilities. Each ophthalmic gadget has a unique IP address; for example, 192.168.1.16 has been given to the fundus capture device, and its associated computer has an IP address of 192.168.1.15. An IP address can be set by following these three simple steps (Control Panel- > Network and Internet- > Network Connections- > Use the following IP address). In this example, we have assigned an IP address of 192.168.1.16. Once the IP address is allotted for that specific gadget, the same address is entered in the web browser of the PC desktop placed near the doctor's slit lamp for activating the screen-sharing relay display from the machine (**Figure 5**).

Total patients								
Month	**Total OPD**	**Inside Cubicle**						**Percentage**
		Slit Lamp	**Lenstar**	**Sirius**	**Fundus**	**NCT**	**OCT**	
2020	19,181	19,101	3185	3185	14,207	13,910	1965	**94.94%**
2021	38,390	38,268	5221	5221	25,108	27,467	4328	**96.50%**
2022	27,564	27,502	4053	4053	12,359	20,242	4120	**95.16%**
Total	85,135	84,871	12,459	12,459	51,674	61,619	10,413	**95.71%**
Month	**Total OPD**	**Outside Cubicle**						
		Synopto Phore	**B-scan**	**Pachy Metry**	**Specular**	**Visual Field**	**IO**	
2020	19,181	30	720	6	9	60	2136	**5.06%**
2021	38,390	150	1330	32	256	218	1838	**3.50%**
2022	27,564	155	959	22	956	290	1300	**4.84%**
Total	85,135	335	3009	60	1221	568	5274	**4.29%**

Table 2.
Total number of ocular examinations at various stations performed inside and outside the cubicle.

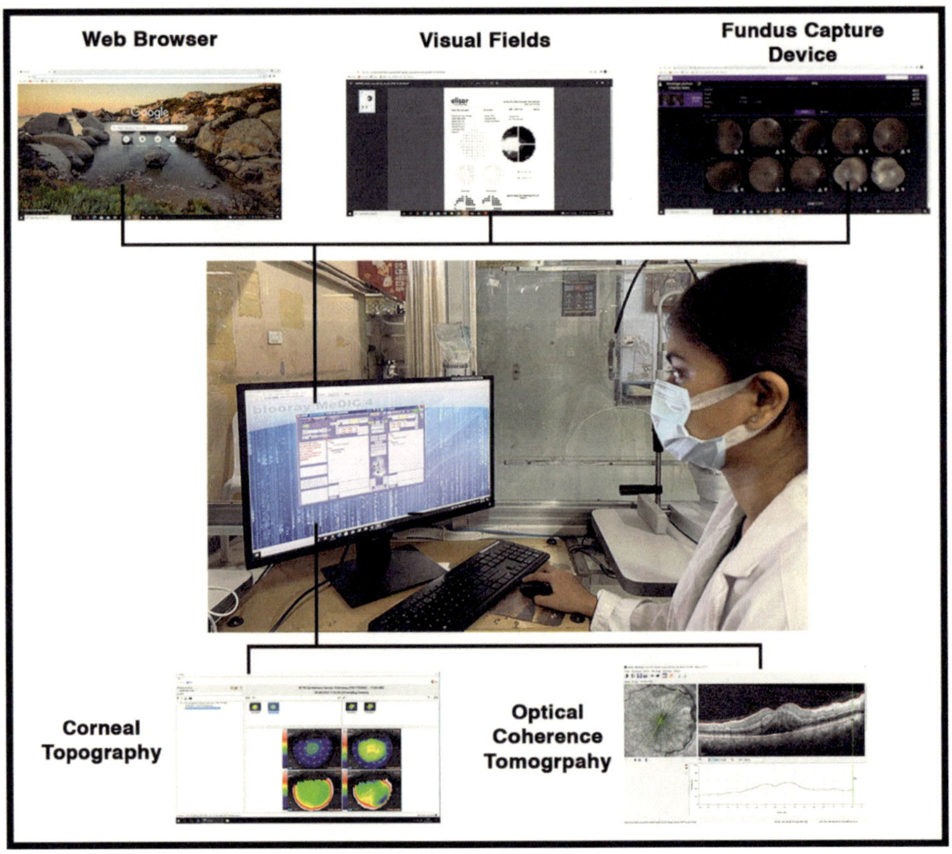

Figure 3.
Image showing various devices linked via local area network (LAN) interconnectivity.

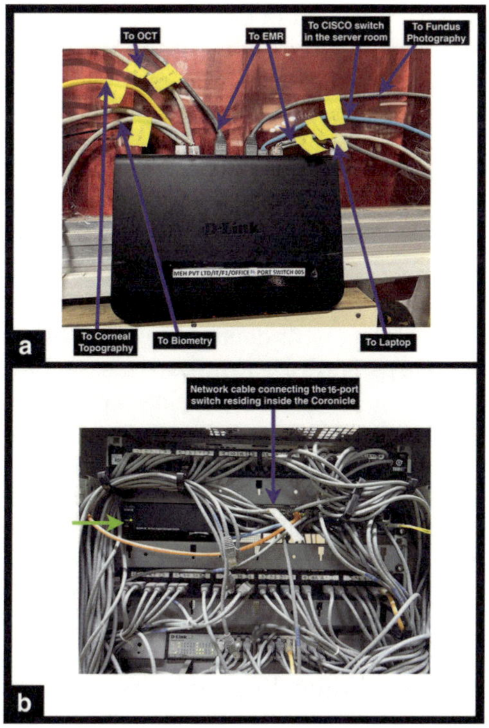

Figure 4.
(a) Sixteen-port switch connections to/from various gadgets. (b) Commercial and industry security corporation (CISCO) switch (yellow arrow indicates the 28 ports in CISCO).

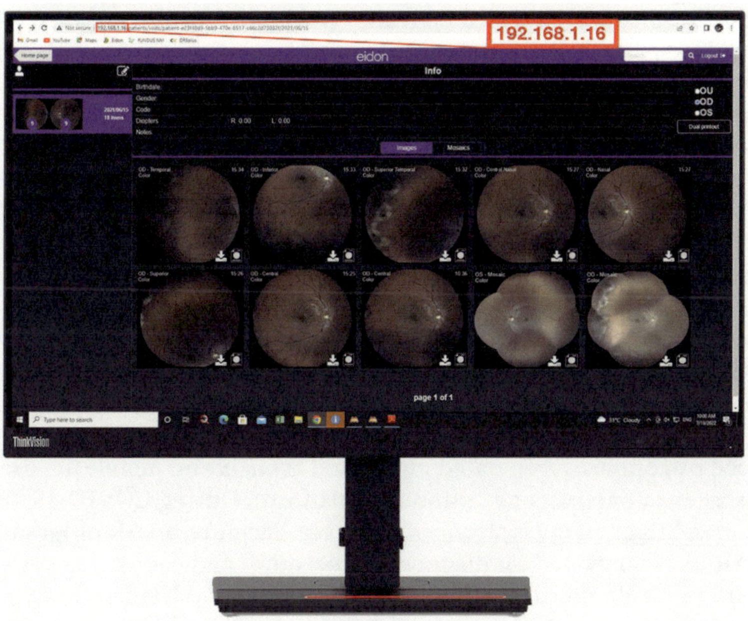

Figure 5.
Image showing LAN connection of fundus capture device after entering the IP address in the web browser.

All eye care professionals can invest in LAN, as it is relatively more cost-effective than a high-end EMR system. The costs of the CISCO server with the switch are approximately 1,00,000 INR/1046.82 GBP and approximately 10,000 INR/104.68 GBP for the sixteen-port switch. The wiring cost depends upon the area covered.

2.2.3 Role of LAN in go green ophthalmic practice

The reports of every patient can be accessed and viewed immediately after the completion of the full evaluation. Also, they can be compared side by side with their former baseline and other examinations. This aids as a smart time-saver. For instance, the time can be reduced considerably, as doctors need not visit each investigative machine to see the respective patient's data and patients can be counseled by showing their image, sitting in one place. At last, the patient's report can be sent by mail or WhatsApp, so they can carry the soft copy for future reference everywhere and anytime. EMR systems require huge storage servers which consume a significant amount of energy, whereas LAN provides better functionality without compromise.

Paperless billing, medical record filing and counseling are now possible with advanced practice management software systems [26]. Going paperless saves trees, which can be considered a direct or indirect way to protect the environment [27, 28].

3. Go green diagnosing and counseling stations

Practicing ophthalmology in a sustainable way begins with a focus on prevention and wellness. Considering the environmental consequences may need more commitment but can be initiated through innovative ideas. There are many creative ways to effectively counsel patients by conserving energy; we have discussed two such innovative ways for ophthalmic patients' counseling and diagnostic procedures. They are three-dimensional (3D) augmented reality (AR) and four-dimensional (4D) holographic diagnosing and counseling platforms. During this fast-paced global era, face-to-face counseling with two-dimensional (2D) images has many challenges, and it is hard for the patients to comprehend their disease. At the same time, the impetus of 3D models in AR and 4D holograms will pay rich dividends (**Figure 6**). The use of such innovations helps reduce the need for conventional paper pamphlets and plastic eyeball models, which makes a significant reduction in the consumption of paper and plastic.

3.1 3D augment reality

AR has progressed from a science-fiction concept to a science-based reality [29–31]. It has slowly but surely become a significant aspect of modern life over the last decade with increasing applications in the field of medicine, especially in ophthalmology [32–34]. AR is a view of the real, physical world in which the elements are enhanced by computer-generated inputs and available on mobile handsets, which constitutes an essential patient e-counseling platform. During COVID-19, the impetus for AR in ophthalmology is stronger than ever. Recently, an AR program named "Eye MG AR" was innovated for diagnostic procedures and counseling patients by showing different anatomical and pathological structures related to the eye (Video 1, https://www.youtube.com/watch?v=LC7VMI56hLo). The patients' own pathological real-time TrueColor confocal images have been used in 3D, with multiple customized angles of the viewer's choice to simplify counseling procedures (**Figure 7**) [35, 36].

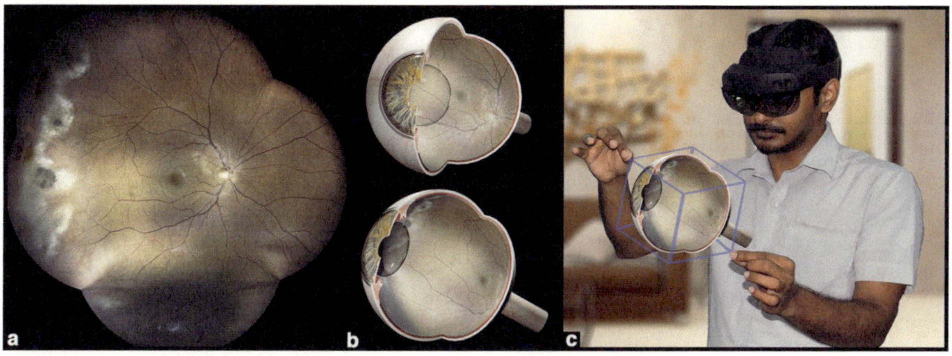

Figure 6.
Image showing (a) two-dimensional (2D), (b) three-dimensional (3D), and (c) four-dimensional (4D) TrueColor confocal fundus images used for patient diagnosis and counseling.

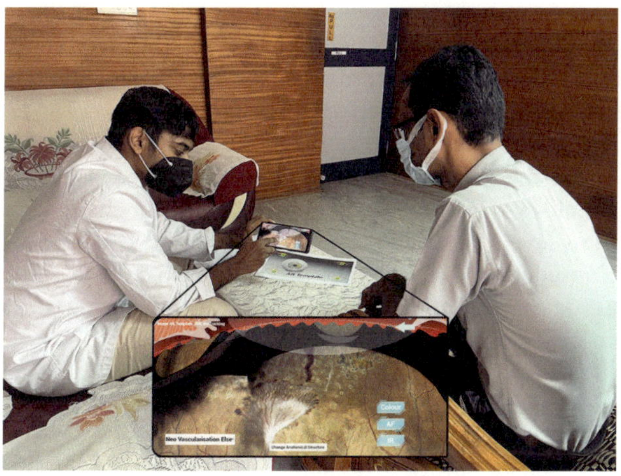

Figure 7.
Image showing the "Eye MG AR" app being used for e-counseling.

For immersive visual experiences, simple structures such as the eyeball and its parts and complex systems essential to the eye (cerebral and dural venous sinuses) were also constructed using advanced real-time 3D photo-real visuals. This app, built on an innovative interactive 3D touch interface, has a significant influence on improving ophthalmic diagnostic procedures, especially in patient counseling.

3.2 4D extended reality holograms

Extended reality (XR) is one of the leading futuristic concepts, which is still slowly evolving to set foot into the field of ophthalmology [37, 38]. What makes this device cutting-edge is the spatial recognition, eye-tracking, and hand-tracking concepts. Spatial recognition senses the world around the user, and eye-tracking recognizes where the user is seeing [39–41]. It also projects the holograms into the eyes of the user as light rays. The hand-tracking concept helps the user to touch, move, rotate, and scale the holograms. Using extended reality technology, especially in ophthalmic diagnostics procedures and counseling, will revolutionize the face of counseling on

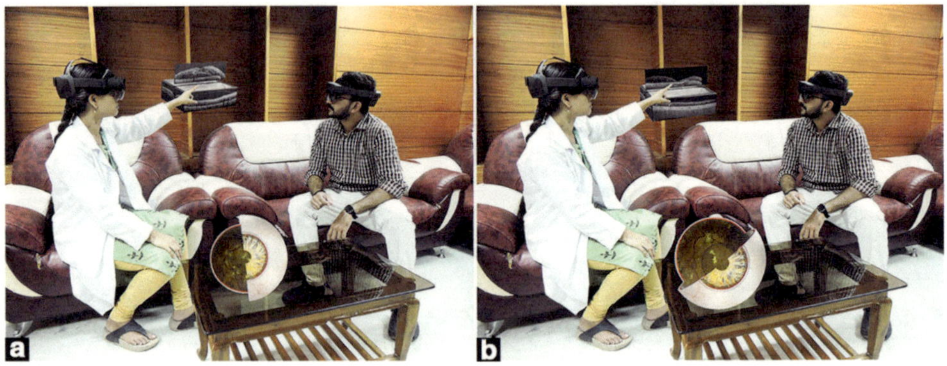

Figure 8.
(a & b) Image showing doctor counseling a diabetic patient with the help of 4D hologram of TrueColor confocal fundus image and optical coherence tomography (OCT) image by using Hololens 2 before and after treatment, respectively.

a whole new level. We have used this novel technology and have created holographic counseling platforms for various anatomical structures such as the eyeball, cerebral venous system, cerebral arterial system, cranial nerves and multiple parts of the brain in fine detail and diseases related to the eye (Video 2, https://www.youtube.com/watch?v=XkHYTzYRHYU). Suppose the patients can see the 4D holographic models related to ophthalmology right in front of their eyes; it can change the way of counseling to a whole new level (**Figure 8**). Ophthalmic institutes and practitioners can invest in this cutting-edge technology to provide their neophyte ophthalmic residents and allied ophthalmic personnel with a real-time understanding of the concepts involved in patient care and diagnosis.

3.3 3D and 4D holographic diagnosing and counseling workflows

The real-time TrueColor confocal images of each patient can be viewed immediately and simultaneously once the patient has completed the advised diagnostic

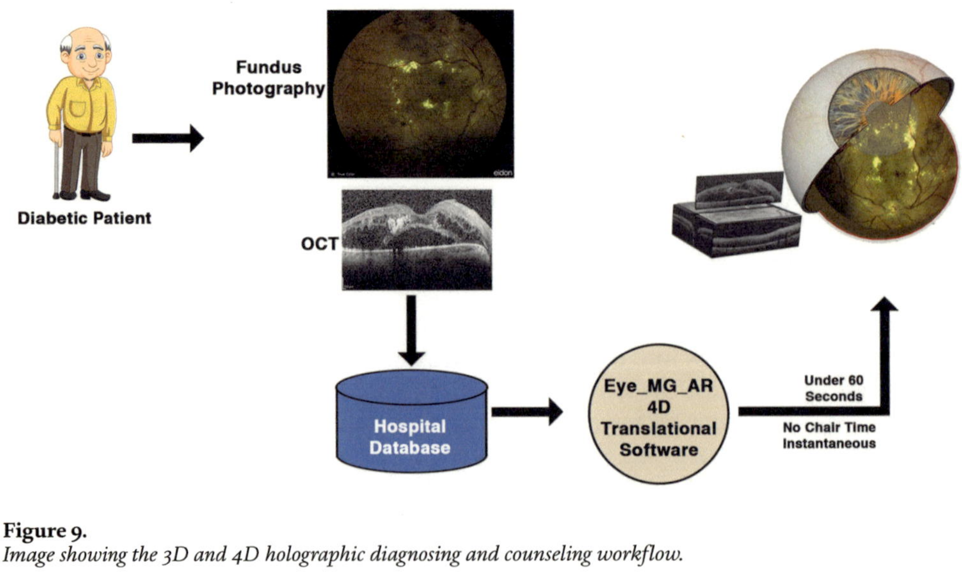

Figure 9.
Image showing the 3D and 4D holographic diagnosing and counseling workflow.

investigations. Also, they can be compared side by side with their previous baseline and other analyses in 4D. This is less time-consuming, eco-friendly and patients can be counseled by showing their pathological images without plastic eyeball models.

After completion of each diagnostic procedure, the patient's data will automatically get stored in the hospital database. Eye MG, the 4D Holographic translational software, will access this hospital database to take the patient's scanned images and project them as 4D holograms (Video 3, https://youtu.be/xZ2b4d89ngk). These holograms can then be used for patient counseling. The whole process will take approximately 60 seconds (**Figure 9**). These holographic case sheets will be stored each time the patient visits. So, for every visit the patient makes after the first time, their condition can easily be matched with their previous hologram side by side.

4. Recommendation

Green practice patterns can start with small steps, which will create an impact in a big way.

- Place recycling bins in convenient locations throughout the hospital.
- Turning off lights and machines at night and setting electronic equipment to sleep mode while not in use saves energy and money.
- Follow the advanced digitalization system for reports to avoid excess paper use.
- Practicing greener surgical protocols to help reduce carbon footprint.
- Ophthalmic institutes and practitioners can invest in 3D AR and 4D holographic technology to provide better patient counseling than plastic eye models without affecting the environment.

5. Conclusion

Keeping in mind that certain current systems in place are not only harmful to the current generation but also have a potential impact on the health of the unborn generation, it is our ethical responsibility to bring change where possible. Though change is a necessity, it is not easy, especially in a large and complex system such as healthcare. So, every small innovation will have a significant role in paving the way toward a green healthcare system. These few easy and adaptable steps of structure-based (cubicle), function-based (LAN), and technology-based (3D and 4D counseling platforms) reforms have been able to pave a path toward sustainable healthcare by helping us reduce our carbon footprint.

Acknowledgements

We are grateful to Mr. Pragash Michael Raj - Department of Multimedia, Mahathma Eye Hospital Private Limited, Trichy, Tamil Nadu, India, for his technical support throughout the making of this chapter.

Conflict of interest

The authors declare no conflict of interest.

Notes/thanks/other declarations

I (Dr Prasanna Venkatesh Ramesh) owe a deep sense of gratitude to my daughters (Pranu and Hasanna) and family (in-laws) for all their prayers, support, and encouragement. Above all, I extend my heartfelt gratitude to all the patients who consented for the images which are utilized for this chapter.

I (Dr. Shruthy Vaishali Ramesh) want to thank my partner (Arul) for his constant support and encouragement during the process of creating this chapter.

I (Ms. Prajnya Ray) would like to offer my special thanks to Mr. Deepak Kumar Panda for his constant support and never-ending encouragement, and my parents (Mr. Anil Ray and Mrs. Soubhagyabati Ray) for their support and motivation during the process of framing this chapter.

I (Mr. Aji Kunnath Devadas) want to thank my parents (Mr. Devadas K and Mrs. Sheeba Devadas) for their constant support and encouragement during the process of creating this chapter.

I (Akshay Surendran) want to thank my parents (Mr. Surendran P and Mrs. Lalitha K) and my sister (Anjana Surendran) for their constant support and encouragement during the process of creating this chapter.

Declaration of patient consent

The authors certify that they have obtained all appropriate patient consent forms. In the form, the patient(s) has/have given his/her/their consent for his/her/their images and other clinical information to be reported in the chapter. The patients understand that their names and initials will not be published and due efforts will be made to conceal their identity, but anonymity cannot be guaranteed.

Financial support and sponsorship

Nil.

Nomenclature

3D	Three-dimensional
4D	Four-dimensional
AR	Augmented Reality
COVID-19	Corona virus disease-19
CFLs	Compact fluorescent lamps
CISCO	Commercial and Industry Security Corporation Switch
DIY	Do-It-Yourself
EMR	Electronic Medical Record

GRIHA	Green Rating for Integrated Habitat Assessment
IGBC	The Indian Green Building Council
LAN	Local Area Network
LED	Light-Emitting Diode
NCT	Non-contact Tonometer
OCT	Optical Coherence Tomography
OPD	Outpatient Department
PC	Personal Computer
POC	Patient Point of Care
UK	United Kingdom
USA	United States of America
XR	Extended Reality

Appendix

3D models were created and used for AR simulations, virtual reality, and advanced mixed reality with Microsoft HoloLens 2 by us in this chapter. This includes the human eyeball with TrueColor confocal fundus image. These models help in better understanding of concepts and can be used in various fields of medicine. We have created models related to ophthalmology, which allows us to explain a disease or a condition with its pathophysiology, pathway, clinical features, tests, treatment and prognosis creatively and interactively.

We have created apps using these models, including the Eye MG AR (https://play.google.com/store/apps/details?id=com.EyeMG_AR) and Eye MG 3D (https://play.google.com/store/apps/details?id=com.EyeMG_3D) which are based on AR model of the eye and multimodal fundus imaging atlas, respectively. These are available for Android users and are free to download from Google Play Store. An app for iPhone users named Eye MG Max (https://apps.apple.com/in/app/eye-mg-max/id1607482649) is currently available in App Store. This application provides eyeball with TrueColor confocal fundus images and all structures related to ophthalmology with a user-friendly interface. In Eye MG Max, multiple views with transparency for viewing the structures passing through another model, free camera mode, annotated modes, customized zoomed views and videos related to any ophthalmic pathology are provided; thus, providing a 3D atlas at the user's fingertip for better patient counseling. Usage of mixed reality in ophthalmology is only in the primitive stages.

The same 3D models used in the previously mentioned applications have been imported into Hololens 2 for holographic viewing by medical professionals and patients. When viewed through Hololens 2, the device tracks our hands, eyes and surrounding environment. This enables us to select different holograms just by looking at them (eye-tracking) and move the holograms around and enlarge/shrink using our hands (hand-tracking). This enables us to view the models in a 360-degree 3-Dimensional view. We can use this to counsel patients in a much more efficient and convincing manner, leading to better patient compliance. We can also add components such as Optical Coherence Tomography apart from TrueColor confocal fundus evaluation. Advancements such as holographic surgical simulative training will come out eventually in the near future.

Author details

Prasanna Venkatesh Ramesh[1*], Shruthy Vaishali Ramesh[2], Prajnya Ray[3], Aji Kunnath Devadas[3], Akshay Surendran[3], Tensingh Joshua[4], Meena Kumari Ramesh[5] and Ramesh Rajasekaran[6]

1 Department of Glaucoma and Research, Mahathma Eye Hospital Private Limited, Trichy, Tamil Nadu, India

2 Department of Cataract and Refractive Surgery, Mahathma Eye Hospital Private Limited, Trichy, Tamil Nadu, India

3 Department of Optometry and Visual Science, Mahathma Eye Hospital Private Limited, Trichy, Tamil Nadu, India

4 Mahathma Ophthalmic Centre of Moving Images, Mahathma Eye Hospital Private Limited, Trichy, Tamil Nadu, India

5 Head of the Department of Cataract and Refractive Surgery, Mahathma Eye Hospital Private Limited, Trichy, Tamil Nadu, India

6 Department of Paediatric and Strabismus, Mahathma Eye Hospital Private Limited, Trichy, Tamil Nadu, India

*Address all correspondence to: email2prajann@gmail.com

References

[1] Costello A, Abbas M, Allen A, et al. Managing the health effects of climate change: Lancet and University College London Institute for Global Health Commission. Lancet. 2009;**373**:1693-1733

[2] Watts N, Amann M, Ayeb-Karlsson S, et al. The lancet countdown on health and climate change: From 25 years of inaction to a global transformation for public health. Lancet. 2018;**391**:581-630

[3] Malik A, Lenzen M, McAlister S, McGain F. The carbon footprint of Australian health care. Lancet Planet Health. 2018;**2**:e27-e35

[4] WHO. WHO Coronavirus (COVID-19) Dashboard. [cited 2022 Sep 20]. Available from: https://covid19.who.int

[5] PNAS. Plastic waste release caused by COVID-19 and its fate in the global ocean, Proceedings of the National Academy of Sciences [Internet]. Available from: https://www.pnas.org/doi/abs/10.1073/pnas.2111530118. [Accessed: June 20, 2022]

[6] Tsai WT. Analysis of medical waste management and impact analysis of COVID-19 on its generation in Taiwan. Waste Management & Research. 2021;**39**(1_suppl):27-33. DOI: 10.1177/0734242X21996803

[7] Buck J. Go Green Initiative. [cited 2022 Sep 20]. Available from: https://gogreeninitiative.org/about/board-of-directors/jill-buck/

[8] Harris N, Pisa L, Talioaga S, et al. Hospitals going green: A holistic view of the issue and the critical role of the nurse leadr. Holistic Nursing Practice. 2009;**23**(2):101-111

[9] Huang MC, Lin JJ. Characteristics and management of infectious industrial waste in Taiwan. Waste Management. 2008;**28**:2220-2228

[10] European Commission Strategies [Internet]. 2017. Available from: http://ec.europa.eu/geninfo/query/action?query_source=RESEARCHPP&swlang=en&QueryText=green. [Accessed: October 25, 2017]

[11] Ravariu C, Mihaiescu DE. Introductory Chapter: Green Electronics Starting from Nanotechnologies and Organic Semiconductors. Green Electronics. IntechOpen; 2018 [cited 2022 Jun 20]. Available from: https://www.intechopen.com/chapters/undefined/state.item.id

[12] Joseph JM, Pyngrope BB, Jose R, et al. Study on awareness of "go green hospital initiative among healthcare personnel with a view to prepare an informational leaflet". Journal of Public Health Policy and Planning. 2019;**3**(2):1-4

[13] Eckelman MJ, Huang K, Lagasse R, Senay E, Dubrow R, Sherman JD. Health care pollution and public health damage in the United States: An update. Health Aff (Millwood). 2020;**39**:2071-2079

[14] Eckelman MJ, Sherman JD, MacNeill AJ. Life cycle environmental emissions and health damages from the Canadian healthcare system: An economic–environmental–epidemiological analysis. PLoS Medicine. 2018;**15**:e1002623

[15] Nansai K, Fry J, Malik A, Takayanagi W, Kondo N. Carbon footprint of Japanese health care services from 2011 to 2015. Resources, Conservation and Recycling. 2020; **152**(2020):104525

[16] Faculty of Public Health Special Interest Group. The NHS: carbon footprint. Available from: https://www.fph.org.uk/media/3126/k9-fph-sig-nhs-carbon-footprint-final.pdf

[17] India among 10 countries accounting for 75% of healthcare's climate footprint. The Indian Express. 2019 [cited 2022 Jun 25]. Available from: https://indianexpress.com/article/india/india-among-10-countries-accounting-for-75-of-healthcares-climate-footprint-5985305/

[18] National Health Service Digital Hospital outpatient activity 2018-19. Available from: https://digital.nhs.uk/data-and-information/publications/statistical/hospital-outpatient-activity/2018-19

[19] J Amy. Health in the Green Economy. Asian Hospital. [cited 2022 Sep 20]. Available from: https://www.asianhhm.com/facilities-operations/health-green-economy

[20] Kwakye G, Brat GA, Makary MA. Green surgical practices for health care. Archives of Surgery. 2011;**146**(2):131-136

[21] Khanna RC, Honavar SG. All eyes on coronavirus – What do we need to know as ophthalmologists. Indian Journal of Ophthalmology. 2020;**68**:549-553

[22] Honavar SG. Prepare or perish – Readiness is the key to reopen for routine eye care. Indian Journal of Ophthalmology. 2020;**68**:677-678

[23] Ramesh PV, Ramesh SV, Ramesh MK, Rajasekaran R. Utilization of hospital car parking garage for COVID-19 triage and screening in a high-volume tertiary eye care center. TNOA J Ophthalmic Sci Res. 2021;**59**:114-116

[24] Ramesh PV, Ramesh SV, Ray P, Devadas AK, Ansar SM, Raj PM, et al. The do-it-yourself (DIY) novel, safe and cost-effective ophthalmic cubicle (coronicle) in COVID-19 era. TNOA J Ophthalmic Sci Res. 2022;**60**:42-47

[25] Ramesh PV, Parthasarathi S, Ramesh SV, Rajasekaran R, Ramesh MK. Interconnecting ophthalmic gadgets (infinity stones) at finger tips (personal computer desktop) with local area network for safe and effective practice during COVID-19 crises. Indian Journal of Ophthalmology. 2021;**69**:449-452

[26] Blum S, Buckland M, Sack TL, Fivenson D. Greening the office: Saving resources, saving money, and educating our patients. International Journal of Women's Dermatology. 2021;**7**(1):112-116

[27] Wong YL, Noor M, James KL, Aslam TM. Ophthalmology going greener: A narrative review. Ophthalmology and Therapy. 2021;**10**(4):845-857

[28] American Academy of Ophthalmology. The Greening of Ophthalmology. 2019. Available from: https://www.aao.org/eyenet/article/the-greening-of-ophthalmology. [Accessed: June 21, 2022]

[29] Gopalakrishnan S, Chouhan Suwalal S, Bhaskaran G. Raman R use of augmented reality technology for improving visual acuity of individuals with low vision. Indian Journal of Ophthalmology. 2020;**68**(1136):42

[30] Iskander M, Ogunsola T, Ramachandran R, McGowan R, Al-Aswad LA. Virtual reality and augmented reality in ophthalmology: A contemporary prospective. Asia Pacific Journal of Ophthalmology (Phila). 2021;**10**(244):52

[31] Karakas AB, Govsa F, Ozer MA, Eraslan C. 3D brain imaging in vascular segmentation of cerebral venous

sinuses. Journal of Digital Imaging. 2019;**32**(314):21

[32] Cabrilo I, Bijlenga P. Schaller K augmented reality in the surgery of cerebral arteriovenous malformations: Technique assessment and considerations. Acta Neurochirurgica. 2014;**156**(1769):74

[33] Eye MG AR - - Apps on Google Play [Internet]. Available from: https://play.google.com/store/apps/details? id=com.EyeMG_AR&hl=en&gl=IN. [Accessed: November 26, 2021]

[34] Aydındoğan G, Kavaklı K, Şahin A, Artal P, Ürey H. Applications of augmented reality in ophthalmology. Biomedical Optics Express. 2020;**12**(5):11-38

[35] Ramesh PV, Aji K, Joshua T, Ramesh SV, Ray P, Raj PM, et al. Immersive photoreal new-age innovative gameful pedagogy for e-ophthalmology with 3D augmented reality. Indian Journal of Ophthalmology. 2022;**70**(1):275-280

[36] Ramesh PV, Ray P, Ramesh SV, Devadas AK, Joshua T, Balamurugan A, et al. Cerebral Arterial Circulation: 3D Augmented Reality Models and 3D Printed Puzzle Models. IntechOpen; 2022 [cited 2022 Mar 14]. Available from: https://www.intechopen.com/online-first/80306

[37] Ong CW, Tan MCJ, Lam M, Koh VTC. Applications of extended reality in ophthalmology: Systematic review. Journal of Medical Internet Research. 2021;**23**(8):e24152

[38] Iskander M, Ogunsola T, Ramachandran R, McGowan R, Al-Aswad LA. Virtual reality and augmented reality in ophthalmology: A contemporary prospective. The Asia-Pacific Journal of Ophthalmology. 2021;**10**(3):244-252

[39] HoloLens 2. Available from: https://medtrixhealthcare.com/holoLens-2blog-post. [Accessed: January 14, 2022]

[40] Sostel. Eye tracking - Mixed Reality. Available from: https://docs.microsoft.com/en-us/windows/mixed-reality/design/eye-tracking. [Accessed: January 14, 2022]

[41] Ramesh PV, Joshua T, Ray P, Devadas AK, Raj PM, Ramesh SV, et al. Holographic elysium of a 4D ophthalmic anatomical and pathological metaverse with extended reality/mixed reality. Indian Journal of Ophthalmology. 2022;**70**:3116-3121

Section 4

Waste Management for Green Computing Systems

Chapter 8

Analysis of Rainwater Harvesting Method for Supply of Potable Water: A Case Study of Gosaba, South 24 Pargana, India

Subhashis Chowdhury, Souvik Chakraborty and Rajashree Lodh

Abstract

In Gosaba, a village on the outskirts of South 24 Parganas, West Bengal, India, people experience a lot of problems related to shortage of potable water due to salinity and arsenic contamination in the supplied water. Rapid growth of industrialization, increased population, saline water intrusion etc. is causing a decrease in fresh water. Due to overuse of groundwater, GWT is declining rapidly in the Gosaba region. Moreover, seawater is intruding into the groundwater, causing pollution of surface water and a rise in Fe content, Cl content, arsenic content and salinity content in groundwater of that location. The runoff available from that amount of received precipitation is estimated using two empirical equations derived by Sir Aiexander Binnie; Ingels-De Souza and T.G. Barlow and the calculation confirms a good amount of runoff that can be utilized for harvesting in order to decrease the water scarcity of the location. The scarcity of fresh water in the Gosaba location can be minimized by adopting the rainwater harvesting (RWH) method, a sustainable process to obtain disinfected water at a very low cost. The technical part of the present study is to adopt RWH where rainwater is collected from rooftop of an institute building and to design tank where water can be stored and utilized further at minimum costs.

Keywords: salinity, groundwater, arsenic content, water harvesting, runoff calculation Gosaba block

1. Introduction

Various regions of West Bengal as well as India are facing the problem of water shortage due to depletion of groundwater table (GWT) at an alarming rate. South 24 Pargana located in the coastal area of West Bengal, India is one of them. Gosaba is one of the blocks in South 24 Pargana. The latitude and longitude of Gosaba are 22.16°N and 88.80°E respectively. The population density of South 24 Pargana is larger than that of West Bengal. Due to excessive use of groundwater, saline water intrusion is taking place rapidly in South 24 Pargana. The groundwater of Gosaba is contaminated by arsenic and also affected by saline water intrusion from the Bay of Bengal. Thus, there is scarcity of freshwater required for domestic and drinking purposes in the Gosaba region. It is very much needed to provide fresh water to different sectors and most importantly for household purposes. Since during monsoon tropical regions receives heavy rainfall, rainwater can be stored for future. Therefore, as an alternative rainwater harvesting (RWH) can be adopted to provide fresh water to the household during dry seasons at a reasonable cost of installation or for recharging groundwater. There are three processes involved in RWH technique: collection of rainwater in a waterproof surface such as impervious ground surface or roof, conveying the water through pipes and conduits from catchment to suitable storage tanks and storing the water properly at some location, may be rooftop or underground tank for future use. It includes primary screening of unwanted materials, first flush diverters and a water treatment plant. The harvested rainwater can decrease the dependency on groundwater caused by increased rate of population. RWH techniques can also prevent soil erosion and flood in coastal areas during rainy season as the excess water is collected in huge tanks. Storage tanks made up of concrete are most widely used in India to store rainwater and can be built below and above the ground surface. Although it is advisable to provide tanks above the ground surface as it is cost- effective and any damage, leaks, cracks etc. can be easily identified and tanks could be cleaned at regular intervals. Moreover, there remains a risk of contamination of water in underground tanks.

Various researchers worked on the RWH techniques to reduce the scarcity of fresh water and save the groundwater. Awawdeh et al. [1] studied that the chronic water shortage in Jordan can be reduced by increasing the amount of rainfall harvesting from rooftops, roads, and parking lots. They conducted a study at Yarmouk University and evaluated the potentiality for potable and non-potable water savings by using rainwater [1]. Tobin et al. [2] studied the practice of rainwater harvesting (RWH) in a rural community in Edo State, Nigeria. A cross-sectional study design and a structured observational checklist were used for the assessment of the household rainwater harvesting system. Data were analyzed using statistical package for social sciences (SPSS) version 16 and results were presented as frequencies. It has been found that RWH was practiced by over 80% of households, with the rooftop as the catchment area. Stored water was most commonly used for personal hygiene purposes [2]. Roy [3] conducted an empirical study on the Bandu river basin in Puruliya district to explain the importance of rainwater harvesting in drought-prone areas of West Bengal. Puruliya district ranks first in vulnerability to drought hazards within the state of West Bengal. The Bandu river basin receives 1150 mm of rainfall even in the driest years but the distribution is uneven, as a resulting scarcity of water occurs. There are 31 villages in the Bandu river basin and the population density of the area is 375 persons/sq.k.m. The objective of the work was to adopt rainwater harvesting and estimate the runoff from the amount of rainfall received. The runoff was estimated through four empirical formulae and the results indicate the amount is a healthy one [3]. Khan [4] made a detailed review of the contribution of rainwater harvesting in the field of agriculture in the Ahmadabad region of Gujarat. Because of the little supply of water in villages by the government, the rural population is fully dependent on water for agricultural as well as domestic purposes. Both primary and secondary data from water samples and community surveys were used to analyze the costs and benefits of RH in the district. The major costs include the initial construction cost of the rainwater harvesting system and the maintenance costs. The major benefits include an increase in household dispensable income, time and energy saved from collecting water, and relief from epidemic droughts [4]. Said [5] carried out a case study in South Delhi to assess the potential of roof-top rainwater harvesting procedure. The study aims to explain rooftop rainwater harvesting that can be easily used by each individual with ease. Data were collected and analyzed in relevance to the actual average annual water consumption of each household and the volume of rainwater collected annually from an individual respective rooftop. The study advised the reduction of 20% in the per capita demand of each individual. The present study finds its usefulness in developing awareness towards the proper use of rainwater for sustainable management of water resources at an individual level [5]. Shittu et al. [6] used a rainwater harvest system to combat perennial water scarcity at the household level in Ibadan city. They collected rainfall data for a period of 10 years. The RWH System used comprised of six basic components: Roof Catchment; Gutters and Downspout; Conveying and Water Treatment, Leaf Screen and Roof-washers; Storage Tanks [6]. Kulkarni [7] summarized studies, research and surveys carried out to study, analyze and implement RWH. It has been observed that the use of the RWH method can fulfill more than 50 percent of water demand in domestic households. The commercial hubs, school complexes, and office premises have more potential for rainwater harvesting [7]. Khan et al. [8] adopted a rainwater harvesting system for the design of optimum rainwater storage tank size and efficiency assessment. The software developed by them proved to be satisfactory for any combination of location, catchment area, material, and water demand and can also estimate the reliability of the corresponding water supply system. The rainfall data for a 24-year period for different areas of Bangladesh was collected from Bangladesh Meteorological Department (BMD) and used in the model. The software was employed to evaluate RWHS in an arsenic affected region (Comilla), coastal location (Khulna), and a low-rainfall area (Rajshahi) [8]. Jagtap and Bhosale [9] developed a working system at a construction site named Daulat Heights in Pune, Maharashtra. The main objective of this paper was to make efficient use of rainwater and adoption of the newly launched concept of nature conversion. The project has to store a capacity of 129,600 liters of water with just Rs 48,060 [9]. Pauline et al. [10] primarily focused on the adoption of water harvesting structures by farmers in dryland areas of Tamil Nadu. A thorough study was conducted and designed by combining a descriptive survey of the study area and a population analysis approach of a participatory study. Different water conservation methods/structures found in the dry areas of Tamil Nadu are farm ponds, tank irrigation systems; compartmental bunding and recharge pit, etc. have been described [10]. Chakraborty et al. [11] have simulated the GWT in various locations of Purba Medinipur with the help of Visual MODFLOW. The lowered GWT is one of the major reasons for salinity in groundwater in Purba Medinipur. From the review of literature, it is evident that RWH technique is advantageous and can be developed for both quantitative and qualitative approaches for the present study area [11]. Chowdhury et al. [12] conducted a comprehensive study in West Bengal and Assam that revealed that there is substantial amount of arsenic affected areas in these two states that is affecting the quality of groundwater and causing various diseases in human being. They also provided few suggestions to treat the arsenic sludge [12].

The present study aims at calculation of runoff available for storing during rainy season using two empirical equations derived by Aiexander Binnie; Ingels-De Souza and T.G. Barlow; determination of Ar content, Cl content, Fe content and salinity in groundwater of South 24 Pargana, designing and installation of RWH plants and explaining the economic benefits of RWH methods.

2. Study area

The study area taken is in Gosaba location that is a block at Jayanagar in South 24 Parganas of West Bengal, India with an area of 297.6 Km^2. It is located in the proximity of the Bay of Bengal. The lithological character of South 24 Parganas is such that it is made up of Entisols, Alfisols, and Aridisols. Horizontal hydraulic conductivity

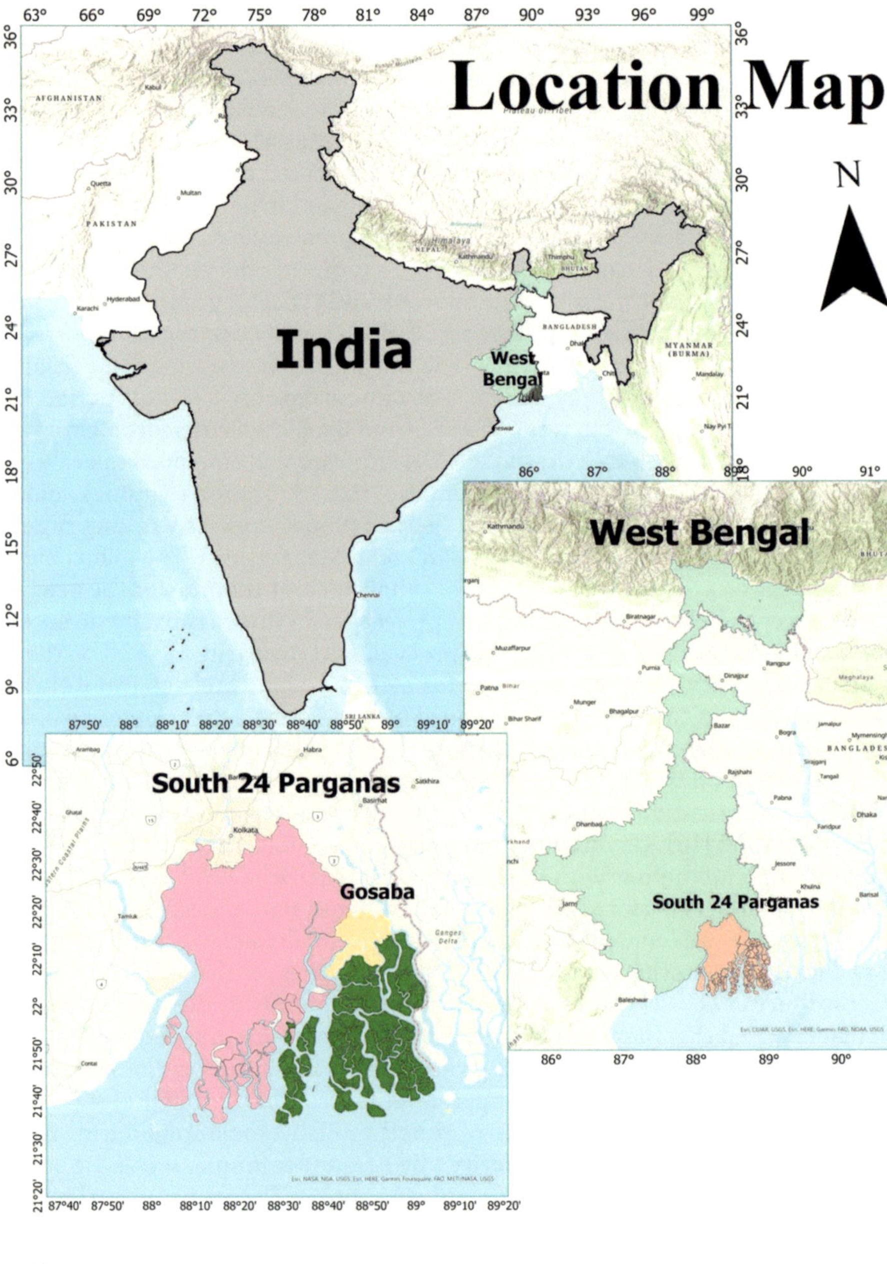

Figure 1.
Study area.

of this type of soil is more than vertical permeability and for that reason; it is more susceptible to seawater intrusion. The study area is represented by **Figure 1**.

3. Status of water sources in the present study location

Groundwater used for domestic water consumption in South 24 Parganas is sixty-eight (68) Million cubic meters (MCM) which is very much significant, as a result, groundwater is depleting very fast. According to Census India 2011 report, the population of Gosaba block was five thousand three hundred and sixty-nine (5369) that has increased substantially in the current year 2021, the increment being around 7% p.a. The population in Gosaba is around 10,000 as of the year 2020. The density of population in South 24 Parganas is more than that of West Bengal as a consequence

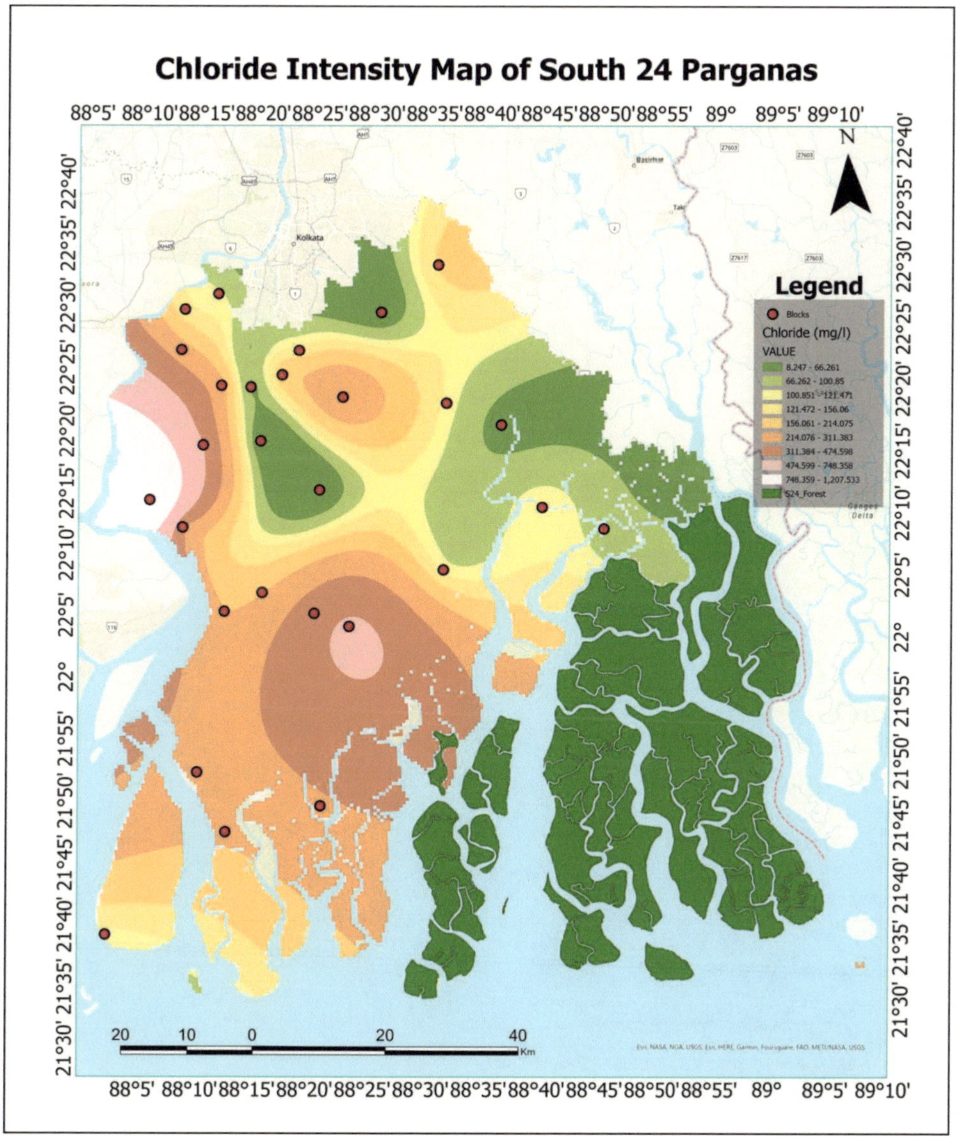

of which groundwater is extracted more resulting in saline water intrusion as per Ghyben Herzberg Principle that states one scale decline of groundwater results in forty scale intrusion of seawater into the mainland. From the information provided by Public Health Engineering Department (PHED), South 24 Pargana, it has been confirmed that groundwater is contaminated with arsenic, arsenic contamination in Jayanagar was 0.05 mg/l, salinity in groundwater in the year 2011 was ranged from 0 ppm at Budge Budge to 4624 ppm in Mathurapur-II and in Jayanagar-I was 2195 ppm. Iron content in groundwater is also a big problem in South 24 Pargana. The concentration of iron in groundwater is ranged from 0 at few places to 6.15 ppm at Magrahat II and in Jayanagar is 3.96 ppm which is beyond the acceptable limit according to IS 10500:2012. The contour map of chloride, iron content and salinity in groundwater of that location is shown in **Figure 2**.

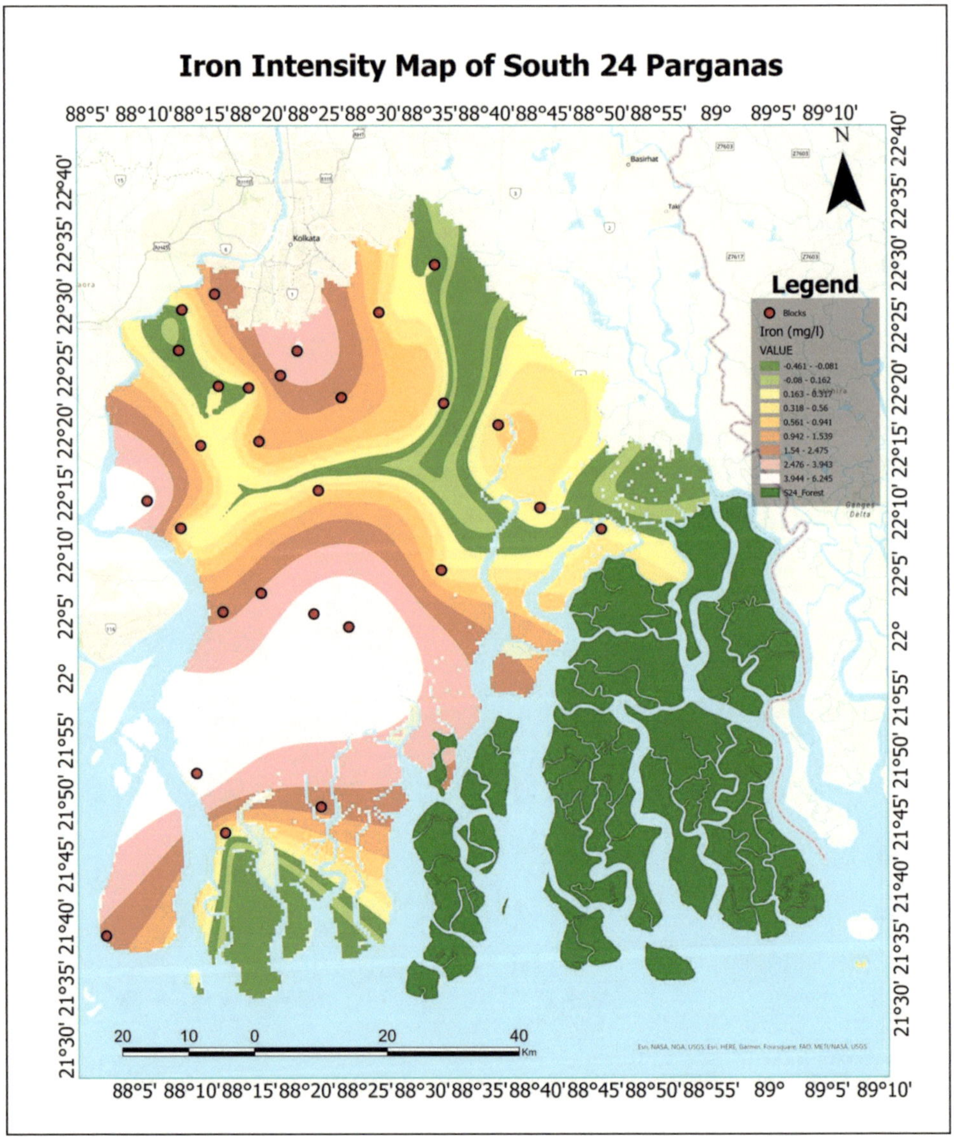

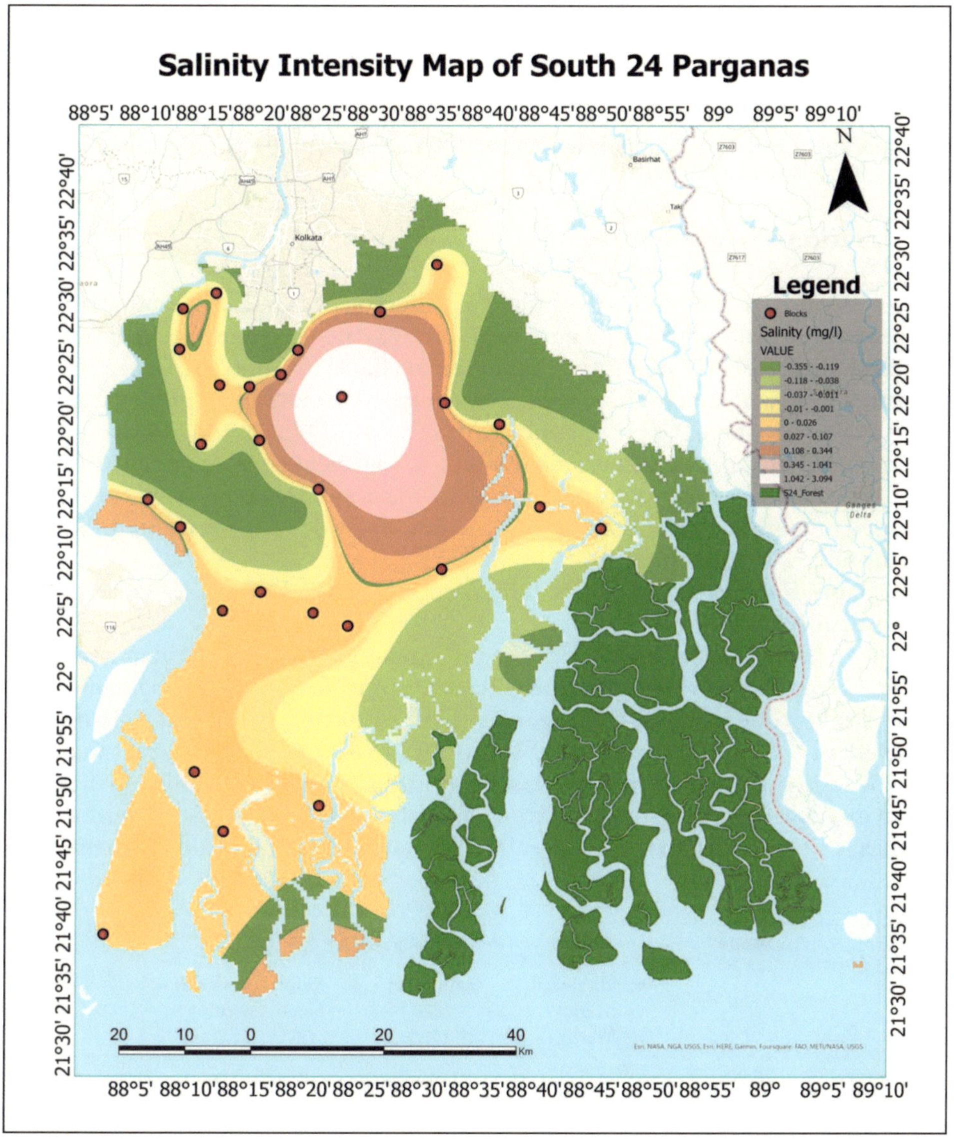

Figure 2.
Groundwater quality parameter's contour map in South 24 Parganas of West Bengal.

Similarly, arsenic contamination in Jayanagar was 0.05 mg/l. Water samples were taken from twenty-nine locations of South 24 Pargana. According to IS 10500:2012, the permissible value of salinity in drinking water should be within 1000 ppm and similarly, the permissible value of arsenic in drinking water should be less than 0.01 mg/l. By inspection of the above value, it is found Gosaba block in Jayanagar was both saline and arsenic affected. Surface water is also saline due to the geographical location of South 24 Pargana. So, Gosaba block in Jayanagar is water scarce area. Water samples were taken from twenty-nine locations of South 24 Pargana. According to IS 10500:2012, the permissible value of salinity and arsenic in drinking water should be within 1000 ppm and less than 0.01 mg/l respectively. After thorough investigation, it has been found that the Gosaba block in Jayanagar is water scarce area as the groundwater is both saline and arsenic affected and surface water is also saline due to the geographical location of South 24 Parganas. Moreover, according to Central Water Management Index, 2018 by NITI AAYOG, South 24 Parganas have been found as water scarce area with the ratio of total water to total population available less than 1000 per cubic meter per capita per year.

4. Estimation of runoff

There are different methods used for determination of runoff in a catchment. In the present study some tables and empirical formulae are used depending on availability of data to estimate the quantity of rainwater available for runoff in Gosaba block.

Binnie's coefficient method: According to this method, runoff and rainfall are correlated as $R = KP$, where P is precipitation in cm, R is runoff in cm and K is the runoff coefficient that changes with the landcover type (Usual values of K are given in **Table 1**).

Ingles and De Souza's formulae: According to Ingels and De Souza, runoff calculation based on rainfall can be done in two separate equations one for plains and the other for hills. The equations are –

For plains, $R = (P\text{-}17.8)\ P/254$; for hilly areas, $R = 0.85P\text{-}30.5$. The calculations are shown in **Table 2**.

Barlow's Table: T.G. Barlow divided the catchment into 5 different types named as A-E after many studies to estimate the percentage of runoff generated by them. The present study area comes under category A (Flat, cultivated, absorbent soil) and the coefficient of this type is 25 percent of the total rainfall. Thus, the equation stands $R = 0.25P$, and the calculation of availability of runoff according to this equation is given in **Table 3**. The seasonal variation of rainfall in Gosaba block of South 24 Parganas is shown in **Figure 3** (Source: IMD Kolkata [15]), that depicts that the area

Type of area	Values of K	Runoff generated by			Calculated average runoff R (cm)
		Normal rainfall in district (196 cm)	Rainfall in 2020 (191.8 cm)	Rainfall in Gosaba block according to CGWB (175 cm)	
Industrial and Commercial Area	0.90	176.40	172.62	157.50	168.84
Concrete or Asphalt Pavement	0.85	166.60	163.03	148.75	159.46
Urban Residential	0.3–0.5	78.40	76.72	70.0	75.04
Parks, Pastures and Farms	0.05–0.3	34.30	33.57	30.63	32.83
Forests area	0.05-0.2	24.50	23.98	21.88	23.45

Table 1.
Average availability of rainwater for runoff (mm) according to Binnie Values of runoff coefficient K (source: Ministry of Water Resources, GoI [13]); Rainfall Data (source: Meteorological Department, GoI [14]).

Land type	Runoff generated by			Calculated average runoff R (cm)
	Normal rainfall in district (196 cm)	Rainfall in 2020 (191.8 cm)	Rainfall in Gosaba block according to CGWB (175 cm)	
Hills	136.10	132.53	118.25	128.96
Plains	137.51	131.39	108.31	125.74

Table 2.
Average availability of rainwater for runoff (mm) according to Ingels and De Souza.

Land type	Runoff generated by			Calculated average runoff R (cm)
	Normal rainfall in district (196 cm)	Rainfall in 2020 (191.8 cm)	Rainfall in Gosaba block according to CGWB (175 cm)	
Total area	49	47.95	43.75	46.90

Table 3.
Average availability of rainwater for runoff (mm) according to Barlow.

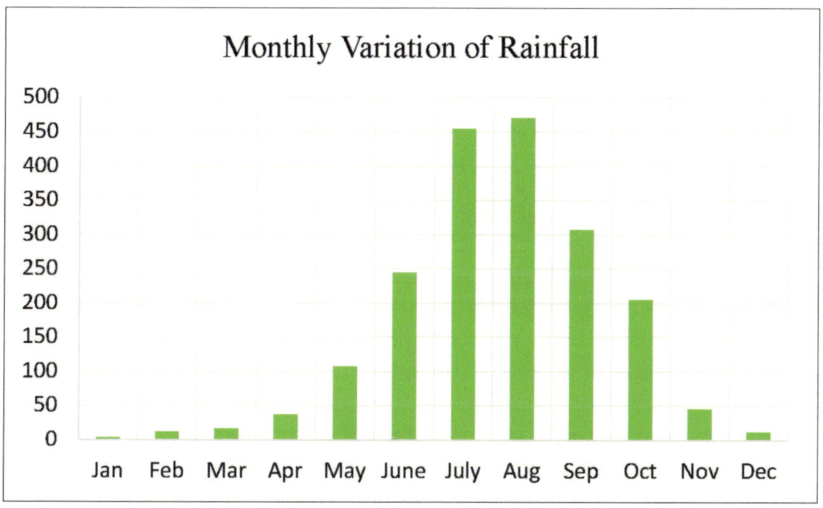

Figure 3.
Monthly variation of rainfall in Gosaba block of S 24 Parganas, WB.

receives maximum rainfall in the month of August and minimum rainfall in the month of January.

From the calculations, it is found that the area receives good quantity of runoff that can be harvested for future use. The runoff needs to be collected, stored and utilized to reduce water scarcity and reduce the overuse of groundwater. During the rainy season excess runoff should be stored for use in dry season. Different amount of runoff is generated by different types of terrain (hilly, plains) and different areas (forest, residential, farms etc.). So, design of RWH plant should be done based on the type of land. The present study area is in plain land, so the calculations and design are done with a normal rainfall of 196 cm.

5. Methodology

5.1 Design of RWH system

- RWH plant is provided at Gosaba R. R. Institute, Bharat Institute of Information Technology, PRITET computer training institute.

- According to IMD, Kolkata information, in 2020 the rainfall in South 24 Parganas was 196 cm. The roof area of Gosaba R.R. College is around 5000 m^2. The coefficient of runoff in this study is taken as 0.95.

- Water from the rooftop of the three institute buildings chosen is stored in the reservoir installed on the ground of dimension (16 m × 8 m × 3.95 m) with 15 cm thickness. The cost of each reservoir is Rs 10,925,400.

- From the rooftop, the water is conveyed through PVC pipe of diameter 75 cm and length 18 m. Specification of the main pipe is 8- inch diameter. The cost of the entire pipe is Rs 1,200,000. For enhancing pressure in low pressure region boost pumping station is also provided. 8-inch pipe is provided throughout the Gosaba block with two pipe branches on either side of the main pipe.

- Treatment units like 2 units of Rapid Sand Filter of length 12 m and width 6 m are provided. The size of gravel, sand used in the RSF are also calculated. The total cost of RSF is Rs 180,000.A rapid sand filter is installed before the reservoir to filter out solid particles in water.

- 3 centrifugal pump sets of 20 hp. with a delivery head of 8 km is provided. Cost of the pump is Rs 90,000 each resulting in a total cost of Rs 270,000.

- In the end, a sufficient chlorine dose is provided to keep the pH range between 6.5-8.5. Freshwater sources in the Gosaba location have been analyzed.

- Volume of rainwater collected in rooftop is calculated as ($V = A \times R \times C$), where R is the rainfall intensity, A is the roof area and C is the runoff coefficient. Considering three RWH plants are installed in the location, the total amount of water stored in a tank in the Gosaba location is calculated as 70 l/capita/day which is much more than the per capita demand of water.

- The total cost of the project serving fresh water to 10,000 people in Gosaba is Rs 35,057,000 which is justified.

The schematic representation of storing to the distribution of rainwater to the consumer has been given below in **Figure 4**.

5.2 Cost analysis of RWH system

The main objective of RWH is to decrease the usage of groundwater and provide both drinking water and non-potable water. The rain water usually contains various contaminants that has to be filtered and disinfected before being used for domestic purposes. Non potable usage of RWH includes gardening, washing and toilet purposes and so requires

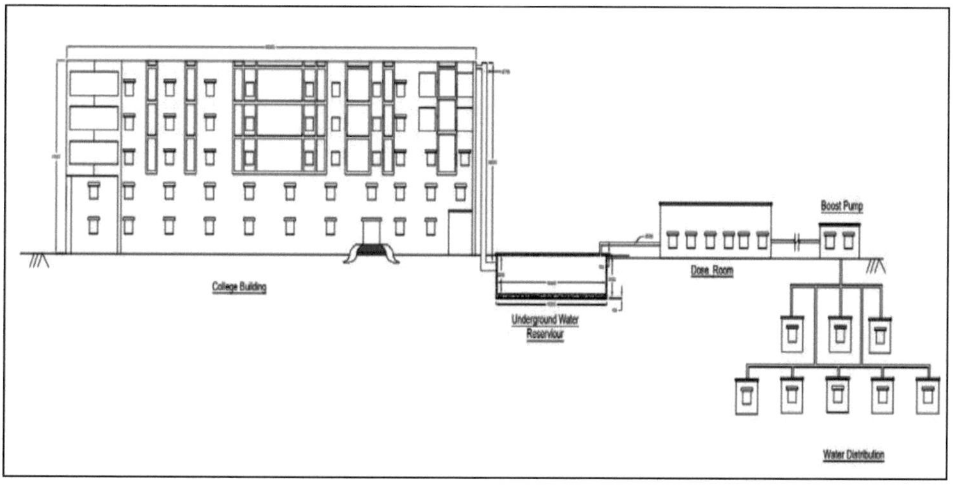

Figure 4.
Schematic diagram of rainwater harvesting scheme in Gosaba, South 24 Parganas.

no treatment. The physical, chemical and biological treatment of harvested rainwater is not required that makes the RWH system economical and cost effective [1]. The storage tank is the costly part of the entire RWH system, the cost analysis presented here consists of cost of materials and construction of storage tank at Gosaba R. R. Institute, Bharat Institute of Information Technology, PRITET computer training institute, South 24 Parganas.

The details of the quantity of materials used for construction of reservoir is given in **Tables 4** and 5.

Sl. No.	Particulars	No.	Length (m)	Breadth (m)	Height (m)	Quantity
1.	Excavations	1	16.30	8.30	3.50	473.51 m^3
2.	PCC (1:4:8)	1	16.30	8.30	0.15	20.29 m^3
3.	RCC (1:1:2)					
	(a) Top slab	1	16.6	8.6	0.15	21.41 m^3
	(b) Base slab		16.30	8.30	0.30	40.58 m^3
	(c) Vertical slab		48	0.30	3.50	50.4 m^3
	(d) Deduction for manhole		$\pi \times (0.25)^2$	–	0.15	−0.0294m^3
						112.56 m^3
4.	20 mm thick plastering (1:2)	1				
	(a) Base	1	16	8	–	128 m^2
	(b) Sides	1	48	–	3	144 m^2
5.	% of steel	1	1/100 × 112.36	1.123 m^3	(1.123 × 7850)	8815.5 Kg

Considering pipe, pump and other ancillary costs total cost of the project is Rs 35,057,000 which is justified.

Table 4.
Details of quantity.

Sl. No.	Particulars	Quantity	Unit/Item rate	Amount (Rs)
1.	Cement	47.75 m^3	280 per bag	3,85,000
2.	Sand	57.3 m^3	58/m^3	3325
3.	Stone chips	105.2 m^3	68/cft	2,52,620
4.	Steel	8815.5	50/Kg	440,775
		Total amount		1,081,720
5.	Water charge	1% of total cost		108,172
		Total amount		1,092,538

Table 5.
Rate analysis of materials.

For concrete work,
Sand = (1/4 × 1.54 × 112.56) =43 m^3; Stone chips = 86 m^3.
For plastering,
Volume of wet mortar = (20/1000 × 272) = 5.44 m^3.
Dry volume of mortar = (5.44 × 1.3) = 7.07 m^3.
Cement = (7.07/3) = 2.35 m^3; Sand = 4.7 m^3.
For PCC work,
Cement = (20.29 × 1.54 × 1/13) = 2.40 m^3.
Sand = 9.6 m^3; Stone chips = 19.2 m^3.

6. Conclusion

The present study assessed the suitability of rainwater harvesting and its use in various domestic and irrigation sectors in a very remote area of West Bengal having scarcity of good quality of water. Gosaba block of West Bengal is a very much water scarce area, as the groundwater in the location is contaminated with saltwater and arsenic. Therefore, there is scarcity of freshwater required for domestic works, irrigation and drinking purposes. In order to get rid of the problem and to maintain the GWT, an alternative method of harvesting rainwater is adopted for the present study. Rainwater harvesting is defined as the method of supplying fresh water in salinity and arsenic affected coastal areas of West Bengal. The availability of rainfall as runoff is estimated based on different empirical equations. The area received maximum rainfall in July–August of 2020 with a normal rainfall of 196 cm. To augment the supply of water to the ten thousand people in the study area, three institutional buildings are chosen. Water is collected at the roof catchment area of those institutional buildings. Through pipe and valve surplus, rainwater is conveyed to the rapid sand filter which is composed of sand, gravel and further carried to the reservoir to store and supply water throughout the year. Before distributing the water, chlorination is applied. By proper grid system and arrangement of boost, station water is supplied to every consumer at an affordable cost of installation of rainwater harvesting system. The tank has storing capacity of 70 l/capita/day of water ad total cost of project is just Rs 35,057,000 which is reasonable. It was also found that the quantity of stored and harvested rainwater could be used throughout the year and not only in rainy season.

Abbreviation

- Ar: Arsenic
- CGWB: Central Ground Water Board
- Cl: Chloride
- Fe: Ferrite
- GWT: Ground Water Table
- IMD: Indian Meteorological Department
- PVC: Poly Vinyl Chloride
- RWH: Rainwater Harvesting

Author details

Subhashis Chowdhury[1], Souvik Chakraborty[2] and Rajashree Lodh[3]*

1 CE Department, Dr. B.C. Roy Engineering College, Durgapur, India

2 CE Department, Dr. Sudhir Chandra Sur Institute of Technology and Sports Complex, India

3 CE Department, Heritage Institute of Technology, Kolkata, India

*Address all correspondence to: rajashree.lodh@heritageit.edu

References

[1] Awawdeh M, Al-Shraideh S, Al-Qudah K, Jaradat R. Rainwater harvesting assessment for a small size urban area in Jordan. International Journal of Water Resources and Environmental Engineering. 2012;**4**(12):415-422

[2] Tobin EA, Ediagbonya TF, Ehidiamen G, Asogun DA. Assessment of rain water harvesting systems in a rural community of Edo State, Nigeria. Journal of Public health and Epidemiology. 2013;**5**(12):479-487

[3] Roy A. Viability of rainwater harvesting in drought prone areas of West Bengal: An empirical study on Bandu River Basin in Puruliya District. The International Journal of Social Sciences and Humanities Invention. 2014;**1**(3):155-164

[4] Khan N. Contribution of rainwater harvesting in agriculture of Gujarat: A case study of Ahmadabad District. IOSR Journal of Economics and Finance. 2014;**5**(5):30-36

[5] Said S. Assessment of roof-top rain water harvesting potential in South Delhi, India: A case study. International Journal of Environmental Research and Development. ISSN 2249-3131. 2014;**4**(2):141-146

[6] Shittu OI, Okareh OT, Coker AO. Development of rainwater harvesting technology for securing domestic water supply in Ibadan, Nigeria. International Research Journal of Engineering Science, Technology and Innovation. 2015;**4**(1):032-037

[7] Kulkarni SJ. Review on studies, research and surveys on rainwater harvesting. International Journal of Research and Review. 2016; **3**(9):6-11

[8] Khan ST, Baksh AA, Papon TI, Ashraf Ali A. Rainwater harvesting system: An approach for optimum tank size design and assessment of efficiency. International Journal of Environmental Science and Development. 2017;**8**(1)

[9] Jagtap YA, Bhosale DP. Rain water harvesting for residential project daulat heights in Saswad. Journal NX—A Multidisciplinary Peer Reviewed Journal. 2018;**4**(4)

[10] Pauline AA, Mahandrakumar K, Karthikeyan C. Adoption of rain water harvesting structures in dry land areas of Tamil Nadu, India. International Journal of Current Microbiology and Applied Sciences. 2020;**9**(3)

[11] Chakraborty S, Maity PK, John B, Das S. Overexploitation of groundwater and as a result sea water intrusion into the coastal aquifer of Egra, Purba Medinipur. Indian Journal of Environment Protection. 2020;**40**(4):413-423

[12] Chowdhury S, Lodh R, Chakraborty S. A comprehensive study on the arsenic contamination in the groundwater of assam and west bengal with a focus on normalization of arsenic-filled sludge from arsenic filters. In: Agarwal P, Mittal M, Ahmed J, Idrees SM, editor. Smart Technologies for Energy and Environmental Sustainability. Green Energy and Technology. Cham: Springer; 2022. DOI: 10.1007/978-3-030-80702-3_13

[13] Central Ground Water Board. Rain Water Harvesting Techniques to

Augment Ground Water. Faridabad: Ministry of Water Resources, Government of India; 2007

[14] Meteorological Department. Report on Rainfall, South 24 Parganas District. Government of India. Kolkata: India Meteorological Department, IMD; 2020

[15] India Meteorological Department (2020) Report on Rainfall, Government of India; 2020

Chapter 9

E-Waste Management in Different Countries: Strategies, Impacts, and Determinants

Shireen Ibrahim Mohammed

Abstract

Over the last two decades, the electronic equipment has increased dramatically around the world, which causes increasing in e-waste as well. This increasing has affected the environment badly. E-waste disposal has become one of the most critical issues and concerns have raised of it because most of these products do not biodegrade easily and they are toxic. Different strategies have been followed in many countries in order to solve the e-waste problem. Understanding these strategies can help to plan better for e-waste management correctly. Awareness of people about the e-waste impacts is crucial, because it can ensure people participation in managing the e waste process. This research has carried out in order to introduce to the e-waste impacts on environment and human health, and the importance of people awareness about these impacts. In addition, it shows many strategies that have been used in different countries to manage the e-waste, choosing the successful one to focus in order to benefit from it. Furthermore, a surveying has been carried out to exam people awareness in Iraq about the e-waste impacts. Finally, recommendations to manage e-waste successfully have been added.

Keywords: e-waste management, e-waste impacts, e-waste disposal, formal and informal recycling, e-waste recycling

1. Introduction

E-waste is informal but popular term. It refers to any electrical or electronic equipment which are in the end of their useful life. Globally, markets of electronic and electrical equipment have grown dramatically. While these products lifespan has become shorter, they are ending up in rubbish dumps or recycling centers. Concerns of the e-waste bad effects have raised around the world. Twenty to fifty percentage of e-waste has been generated per year around the world [1]. It has become a big threat on environment and human health [2]. The major reasons of e-waste growing amount are: Short lifespan of the electronic products [1], growth of population, economic development, and consumption patterns changing [3]. Many countries have followed different approaches to solve the e- waste issue. This research has shown some of these strategies and the successful one has been chosen to focus in order to benefit from it. In addition, a surveying has done to exam the awareness of people about the e-waste effects on their health and environment. Furthermore, recommendations have been added to manage the e- waste successfully.

2. E-waste impacts

E-waste can be classified according to their physical and chemical constituent. The compositions of e-waste include: Metals such as [Copper, Iron, Tin, Nickel, Lead, Aluminum, Zinc, Silver, Gold, and Palladium], Plastics, Metal-plastic mixture, Cables, Screens (CRT and LCD), PCB, Pollutants, Wood, and Refractory and Oxides...etc. [4].

E- waste contaminants can be classified into three types [4]:

- Primary contaminants such as the heavy metals and halogenated compounds.
- Secondary contaminants, which produced by products or residues that are produced as a result of the improper recycling process such as poly aromatic hydrocarbons (PAHs), dioxins, and poly halogenated aromatic hydrocarbons (PHAHs).
- Tertiary emissions or contaminants: Compounds which are used for recycling. These compounds must be handled properly in order to avoid bad impacts on environment and health such as aquaregia, nitric acid, cyanide, hydrochloric acid, thiourea, and bromide in the metal recycling leaching process [5].

The unsuitable recycling activities of e-waste such as open burning and manual dismantling cause soil and river pollution. This is discovered by finding fire retardants in soil and river sediments in Vietnam [6]. Soil pollution also was found in china near the e waste recycling area [7]. By observing the heavy metal concentration in the air, it found that in e-waste open burning site the levels of these metals are higher, and the air pollution in these areas was higher [8]. Also it was found that in e waste area that the PCBs and BFRs levels in indoor dust were higher than non-e-waste area [9]. In addition, it was discovered that the illegal dismantling of e-waste with open-air burning have bad impacts on the groundwater [10], and surface water [11].

Several studies have indicated that e-waste has bad effects on the human health. By testing the blood, hair and urine, it have discovered that people who live near the e waste sites have high $\sum$mPAEs concentration in their urine than other people who live in non e waste site [12]. In addition, carcinogenic metabolites has exist in the respondents internal hair [13]. Furthermore, exposure to the heavy metals has caused acute and chronic effects such as respiratory reproductive problems, cardiovascular, and irritation [14]. Studies have shown by testing the DNA of workers who recycling e-waste, there is a correlation between damage of DNA and duration of e waste processing in informal e-waste recycling site [15–17]. E-waste also causes spontaneous abortions, premature births, and reduced birth length [18].

3. E-waste recycling

Recycling the e-waste seems to be good solution in many countries as long as a valuable items can be extracted from it. E-waste recycling is very important for environment sustainability and for economic recovery. The efficient recycling of electronic scrap has been regarded as a major challenge for many countries.

Recycling is the most important key to reduce the e-waste. It has environmental benefits at all stage in the life cycle of the electronic products, from the raw materials from which they are made to their final methods of disposal. Recycling also contributes in reducing water and air pollution which is associated with creating new equipment from raw materials [1]. Generally, people in some countries realize that there is a value generating from different types of household solid waste [19]. So informal peddlers and formal collectors pay to consumers for their waste. Then they sell to refurbishes, brokers, scrap dealers, and recyclers.

Recycling in the world can be divided into two types: Informal recycling and formal recycling.

3.1 Informal recycling

The informal recycling of e-waste consider as a way to extract value from waste electrical and electronic equipment. Informal sector is illegal because it is outside of official institutions [19]. The majority of the informal recyclers are from rural areas and most of them are women and children [20]. This type of recycling includes labor intensive and dangerous manual dismantling of equipment. In this kind of recycling, simple tools are used such as chisels, hammers, and screwdrivers in order to achieve swift separation of the different materials [21]. This sector uses substandard processes and does not have the appropriate facilities to safeguard human health and environment, so it causes risk for recyclers and for environment [19–21]. Informal recycling sector has grown in many countries such as china, Bangalore, Chennai, India, Nigeria, and Pakistan. Informal recycling effects can be shown in **Figure 1**.

There are several reasons for informal recycling increasing [19–21] include

- Awareness lacking of collectors, recyclers and consumers of dangerous improper of e- waste handling.
- Appropriate management absence of e-waste recycling.
- No effective take-back programs for end –of-life electronic equipment and obsolete.
- Absence of interesting in e-waste managements by IT companies.

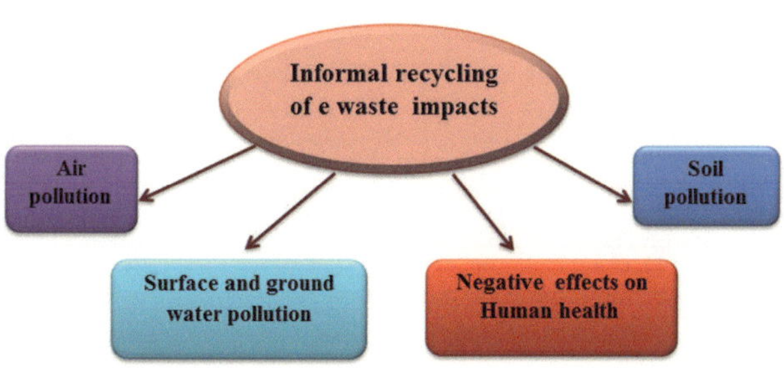

Figure 1.
The effects of informal recycling.

- Implementation laxing in e-waste specific legislation.
- Most recyclers are from rural areas and most of them are women and children.

3.2 Formal recycling

Formal e-waste recyclers can be defined as all companies designated recycling that are included on the e-waste dismantling enterprise list and have a treatment license, which are issued by provincial environmental protection bureaus [20, 22]. Formal sector recycling process are safer for both workers and environment, because it is controlled by the governmental regulation and financing [22]. However, formal recyclers have to bear all the cost (collecting, transporting and disposing of hazardous fractions), while informal recyclers bear less expensive recycling practices. This is because informal recyclers benefit from the reduced costs in terms of recycling technology, collection, pollution, and control systems [23].

4. E-waste management system in different countries

Countries are divided into developed countries and developing countries.

4.1 E-waste management system in developed countries

The best e-waste management can be found in European union. For example, best e-waste management systems can be found in Switzerland and the Netherlands [1, 24]. The European Union has a law for requiring companies and manufacturers to prepare to disposals from e-waste by special mechanisms [1]. It has restricted the use of certain hazardous substances in electronic and electrical equipment. In addition, they put a law for the WEEE companies to set up systems for the WEEE treatment and producers. This system is to be responsible about their products over the entire

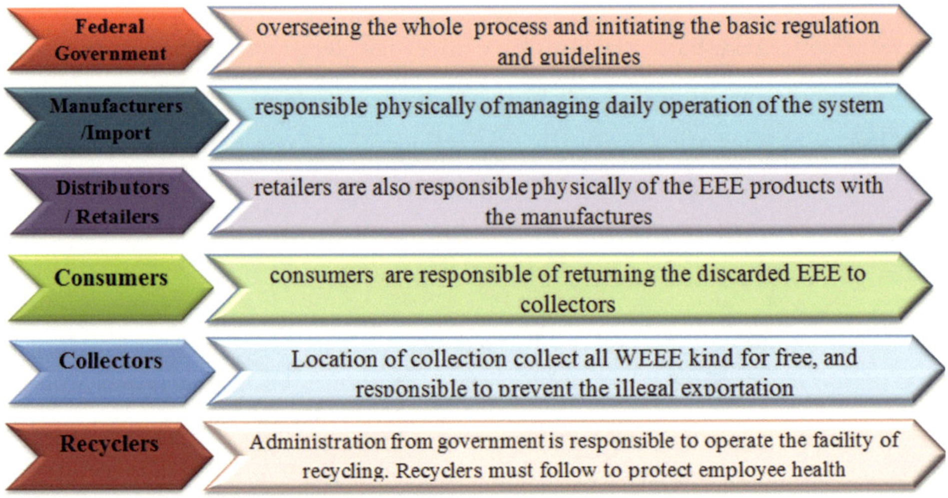

Figure 2.
E-waste management system in Switzerland [1, 24].

lifecycle of their products from design to use till disposal. In other words, Extended Producer Responsibility (EPR), which can be defined as "the producer's responsibility for a product is extended to the postconsumer stage of a product's life cycle" [1, 24].

As far as Switzerland is concerned, it has long experience for applying the EPR for e waste management. In Switzerland every sector has its own role and responsibility as shown in **Figure 2.**

In Switzerland, there are three producers responsibility organizations (PROs) (The Swiss Association for Information, Communication and Organizational Technology (SWICO recycling), the Swiss foundation for waste management (SENS), and the Swiss lighting recycling foundation (SLRS) [25]. These organizations are responsible of the e-waste management. The EPR system has initiated by EEE producers voluntarily in order to manage e-waste. This system has many advantages include [22, 25, 26]:

- its clear system, each one has its own responsibilities of all actors.
- by implementation the advance recycling fee ARF principle, the financing is done.
- accepted and comprehended system and covers wide range of EEE
- control the mechanisms exist for emissions and health hazards.
- monitoring the financial and material flows
- prevent the illegal e waste dumping to the non-OECD countries for recycling
- producers of EEE had Initiated an organized system to manage the e-waste. For instance, reducing the CFCs' emissions from refrigerators and air conditioners.
- The not-for-profit SENS system has managed by members, are the importers, manufacturers, distributors and wholesalers.
- Consumers have to dispose the e-waste through retailers or by collecting it at a located collection points, and they have to pay the ARF for purchasing a new product. So they bear the responsibility of the final financial for e-waste management.
- The recyclers are responsible for environmental sound management of the e-waste. Their focusing is on the efficient materials recovery with reducing the final remains which end to the landfills or to the incinerator. E-waste then is sorted into categories and invoices are send to SWICO and SENS accordingly.
- recyclers must adhere to the standards of emission and safety measures.
- Recyclers operate recycling facility that is authorized from the cantonal government and licensed by the PROs. Government takes the overseer role without involving in day-to-day activities.
- The federal government contributes in guidelines framing for the major involved actors and assigns the key legal responsibilities to them.

- Switzerland cantonal government has the power for issuing and revoking permits of recyclers.

- Disposers are the incineration and landfill facilities, who receive the e-waste processes remaining. They have to follow the safety norms. un-processed e-waste is not allowed to be in the landfilling. in Switzerland, land filling of is completely banned [22].

E-waste in other developed countries nearly the same with some different. In Germany, strict rules have set in order to manage the e-waste. The e-waste is collected directly from individual households without any charging from consumers for e waste disposing, and the collecting points numbers are depending on the population density and local conditions [22]. Informal collection is not allowed. After collection stage, e-waste are handover to the producers. Then e-waste is separated to five different containers. They divided to five according to the categories that are specified in the Act [22]. In the US, The US disposes from e-Waste in developing countries, which causes problems in environment and health in these regions [1]. In Japan, the Ministries of the Environment, Economy, Industry, and Trade have enacted the Basic Law in order to promote recycling and conserve resources to tackle with landfill capacity issues and resource scarcity. The main purpose of this law is to establish recycling-based society, minimize the e-waste generation, and maximize the use of secondary materials [27]. E waste Infrastructure management includes: collecting, logistics and reprocessing technologies. The retailers responsibility is collecting the EoL products from the household to the regional aggregation stations. Also there are many associations to collect the electric home appliances which are appointed by the government. There is end-user-pays principle which makes consumer pay for cost. The consumers can buy a recycling ticket in order to give it to the agent of collection while they discard their e-waste. For instance, if consumer wants to discard his computer, he should contact the manufacturer or he can take it to the post office. Then, it is routed to be recycled in the recycling facility of the manufacturer [28].

4.2 E waste management system in developing countries

In developing countries, there is no specific legislation to deal with e-waste. Developing countries do not have the required infrastructure and technical capacities for waste removal in safe ways, which has caused health problems in these countries such as neurological and respiratory, cancer, disorders, and birth defects [29]. So it has become crucial to prevent the illegal imports of WEEE. In some cases, recycling cost exceeds the revenue that is recovered from materials especially in countries which have strict environment regulations. So the end of e-waste are dumped in countries where standards of environmental are low or nonexistent including Asia and West Africa [29]. Basel Convention which is an international accord, has prevented the exportation of hazardous waste to the poor countries since 1992. The exportation is continued to export what they called it "second-hand goods" as long as the exportation of reusable goods is allowed. But in reality EU Commission estimated that nearly 75% of the second hand goods are broken and cannot be used again or just have short second life to work [1, 29]. E-waste importing from developed countries has found in developing countries because it's considered as a type of livelihood by the residents. Absence of formal recycling in developing countries, makes the e waste and informal waste treatment thrived near the residential areas [4]. Informal recycling processes and informal treatment are done without any knowledge about the hazard that affects the environment and human health in some developing countries such as Indonesia and Cambodia who have not specified law to manage the e-waste [30, 31]. Lacking of data about the amount of material flow of e-waste and lacking of awareness, have affected on the e-waste management in developing countries [32, 33]. Because of the political challenges, technological, and financial, most of developing and under-developed countries are unable to manage the e waste in a way that does not affect the environment and human health [34].

in Africa there is an ineffective infrastructure of the e-waste management. There is no constant system for e waste collection, separation, storage, sorting, and disposal of e-waste. In addition, there is nearly no enforcement effective to regulate the e-waste management and disposal. In Africa e-waste management is reregulation, and rudimentary [35]. Recycling includes disassembly of WEEE without taking into account the hazardous chemicals. For instant, printed circuit boards (PCBs) are heating to recover chips, plastics are melted and burned to isolate metals, this burning sends dioxin and other toxic gases into the air which causes pollution to the environmental and human health. While parts that are dumped in landfills, allows the remained heavy metals to harm the area and life. So, in order to protect public health and the environment, the National Environmental Management Waste Bill in South Africa has implemented for reforming waste management legislation [1].

In Bangladesh the most popular method of e waste management is the dumping into landfills, a small amount of e-waste are recycled. Fresh drives have initiated by policymakers in order to increase the disposal of e-waste, one of these policies is adopting stringent for e-waste management policy [34]. Bangladesh is responsible of 7% of e-waste dumping annually all over the world [36]. Different types from of e waste are produced in Bangladesh every year. it produced from different sources such as mobile phones, televisions....etc. The majority of these wastes are dumped in landfilling or in open water [36].

E-waste management system in Bangladesh can be divided into three categories include:

- Reuse: reselling the e waste after repairing the used electronic and products [37].
- Recycling: Recycling of e-waste is necessary but it is not common process across the country [38].
- Dumping and Landfilling: most of electronics which include computers and mobile phones are disposed of in the litter bins. While medical wastes are burned [34].

In India, only 2 percent of India's total e-waste are recycled because of the lack of legislation and poor infrastructure. This has led to a waste of the diminishing natural resources [39]. In India, several stakeholders are involved in managing the e-waste, so it originates from many sources and does not follow single set path. Most of the e-waste end up with scrap dealers and traders for economic benefits, which end to unorganized sectors [22]. Problems that faces the e-waste management system are lack information about flow and quantum of e-waste, poor infrastructure, poor implementation rules of e-waste, and producers shirking of proEPR [40].

As far as China is concerned, China has made good efforts in order to have better collection and recycling of e-waste in public and private sectors [41]. China's government has issued many of environmental laws and technical guidance related to e-waste management. The most important include [20]:-

- cataloging the import of waste including: the second- hand electronic equipment and e-waste in a list of prohibited imported goods for processing or trade.

- prevent pollution and control the WEEE. This is done by providing a list for environmental measurement to minimize the pollution during the storage, recycling, and final disposal of e-waste.

- prevent and control the pollution caused by electronic and information products. The purpose of this policy is to reduce the using of hazardous and toxic substances in electronic appliances in order to reduce the generated pollution in the manufacture, and recycling of these products.

- Administrative measures to prevent pollution caused from waste electrical and electronic equipment. This is to prevent pollution caused by the transport, storage, recycling, disassembly, and disposal of e-waste. This policy applies for companies who need treatment licenses that is confirmed by local environmental departments...etc.

5. Methodology and result

E-waste problem in Iraq does not different from the situation in other developing country. The problem is the same including: There is no specific legislation to deal with the e-waste, the required infrastructure and technical capacities for e-waste removal in safe ways are not available, experience lacking in recycling the e-waste, and lacking of data about the amount of material flow of e-waste. All these reasons have affected on the e-waste management in Iraq. So the methodology in this research has only focused on the awareness of people in Iraq about the e-waste impacts.

E-waste impacts awareness of people in Iraq from different generations was measured as shown in **Table 1**. The samples of 500 people were chosen randomly to answer the questions. The three points scale questionnaire: 1 = Yes, 2 = No, 3 = Do not know or Not sure. Except the last question which was about the end of their end life mobiles and laptops in specific.

No.	Questions	Yes = 1%	No = 2%	Not certain or Do not know = 3%
1	Do you know what the e waste meaning before the explanation	80	15	5
2	Do you know the negative impacts of e waste on people health and environment	20	35	45
3	Do you with Recycle and reusing different types of e waste	89	3	8
4	Do you separate the e-waste to be recycled later	10	90	0
5	Do you need to enhance your knowledge about the e-waste effects	97	0	3
6	What do you do with your end life mobile or laptop?			

Table 1.
E-waste impacts awareness in Iraq.

Note: This surveying has carried out from 25 to 5-2022 to 10-6-2022.

The result shows that high percentage of people know what is the e-waste meaning, but only 20% of them aware about the negative effects of it on environment and human health. Although the majority of people are with recycling idea and reuse of e waste, only small percentage separate it. Most people admit that they have to enhance their knowledge about the e-waste effects. The last question was asked to have knowledge about the end of their end- life mobiles and laptops in specific. Their answers were different, but they can be divided into three answers: some of them damage their end-life devices and threw in the rubbish. Few of them said that they give their end-life devices to the mobiles sellers or laptop sellers to benefit from the health parts of it. Other mentioned that their end-life devices ended to be with informal collector to make use of them.

6. Discussion and recommendations

According to the result of this surveying, people need to know the causes of environment pollution and its impacts on their health. They need to understand more about e-waste effects and e-waste managements. These two terms are connected with each other, they affect the human health and environment weather in negative or positive way.

Below are some recommendation to manage e-waste successfully:-

- An aggressive legislation must be taken for new technological solutions. Evaluating the present laws and making the suitable modification periodically.

- Increasing the public awareness through education on e-waste and recycling. This is can be done by educating people about how to recycle and dispose electronics and teach them the right behavior to become more responsible towards environment.

- it is the government duty to provide the infrastructure for formal e waste treatment and recycling and encouraging the EEE producers in order to focus on the Extended Producer Responsibility EPR [42].

- In order to have successful formal recycling plant: technical support, guidelines, and financial support, must be exist for treatment processes improvement [43].

- Reuse can gives higher environmental advantages and consider as a safer term in collection of e-waste.

- guidelines must be suggested about the treatment options and technologies to manage the e waste.

- units for recycling of waste EEE must be established.

- consumers must be aware about the hazardous components in products providing with instruction about handling the equipment after its use to prevent e-waste from discarding in garbage bins with other rubbish, stringent provisions are needed

- restricting the second-hand electronics importing, prohibiting the old or used electronic products importing, except electronics that are used for research purposes and in academic institutions with allowable certification from Environment department

- The best example about the e-waste treatment is changing it from primitive practices to developed system such as the pyro metallurgical or hydrometallurgical as occurring in China [4].

- each improvement in each system can be reduce the negative effects of e waste treatment [44].

7. Conclusion

E- waste increases every year at an alarming rate, it is a global problem and needs global solution. So efforts must be taken to minimize illegal dumping. Most developed countries have developed legislations and policy guidelines to control the hazardous chemicals using in these products, and to manage the e-waste after discardation. The European Union has successes in implementing a uniform legislation for e-waste management. Switzerland led the way to establish a successful and formal e-waste management system followed by Germany and Japan.

As far as developing countries are concerned, there are many legislations are existed in developing countries, but they do not come together with enforcement. Most developing countries do not have the similar regulations particularly in their enforcement. In addition, special centers for e-waste processing are not available, facilities scarcity to extract the precious metals from e-waste are not available too. An accurate data of e-waste flows is not available which is a major required for e-waste management system.

By connecting the local and the national regulations, improvement in e waste management can be achieved in developing countries.

People behavior influences e-waste management. Behavior is identified as one of the important factors which affect the e-waste management funding [45]. Awareness of people can ensure that consumers participate in managing the e waste process. So the methodology in this research has focused on people awareness of the e waste. It is the responsibility of government to notify people of their duties and make them aware about e-waste impacts on health and environment due to unsuitable disposal of e-waste. Awareness lacking leads to engagement lacking in re-use and recycling inefficient product use, and lacking of engagement from consumer end. E-waste can be a great energy source if it treated appropriately.

Author details

Shireen Ibrahim Mohammed
Dams and Water Resource Department, College of Engineer, University of Anbar, Anbar, Iraq

*Address all correspondence to: shireenmohammed@uoanbar.edu.iq

References

[1] Khurrum SM, Omar A, Yang X. Electronic waste: A growing concern in today's environment. Hindawi Publishing Corporation Economics Research International Volume. 2011;**2011**:474230. DOI: 10.1155/2011/474230

[2] Woodell D. GeoPedia. National Geographic [Internet]. 2008. Available from: http://ngm.nationalgeographic.com/geopedia/E-Waste

[3] Alamerew YA, Brissaud D. Modelling and assessment of product recovery strategies through systems dynamics. Procedia CIRP. 2018;**69**:822-826

[4] Halim L, Suharyanti Y. E-waste: Current research and future perspective on developing countries. International Journal of Industrial Engineering and Engineering Management (IJIEEM). 2019; 1(2):25-42. DOI: 10.24002/ijieem.v1i2.3214

[5] Khanna R, Cayumil R, Mukherjee PS, Sahajwalla V. A novel recycling approach for transforming waste printed circuit boards into a material resource. Procedia Environmental Sciences. 2014;**21**:42-54

[6] Someya M, Suzuki G, Ionas AC, Tue NM, Xu F, Matsukami H, et al. Occurrence of emerging flame retardants from e-waste recycling activities in the northern part of Vietnam. Emerging Contaminants. 2016;**2**:58-65

[7] Uchida N, Matsukami H, Someya M, Tue NM, Tuyen LH, Viet PH, et al. Hazardous metals emissions from e-waste-processing sites in a village in northern Vietnam. Emerging Contaminants. 2018;**4**:11-21

[8] Gangwar C, Choudhari R, Chauhan A, Kumar A, Singh A, Tripathi A. Assessment of air pollution caused by illegal e-waste burning to evaluate the human health risk. Environment International. 2019;**125**:191-199

[9] Tue NM, Takahashi S, Suzuki G, Isobe T, Viet PH, Kobara Y, et al. Contamination of indoor dust and air by polychlorinated biphenyls and brominated flame retardants and relevance of non-dietary exposure in Vietnamese informal e-waste recycling sites. Environment International. 2013;**51**:160-167

[10] Idrees N, Tabassum B, Abd_Allah EF, Hasehm A, Sarah R, Hashim M. Groundwater contamination with cadmium concentrations in some West U.P. Regions, India. Saudi Journal of Biological Sciences. 2018;**25**:1365-1368

[11] Liu Y, Tang B, Lup X, Mai B, Covaci A, Poma G. Occurrence, bio magnification and maternal transfer of legacy and emerging organophosphorus flame retardants and plasticizers in water snake from an e-waste site. Environment International. 2019;**133**:105240

[12] Zhang B, Zhang T, Duan Y, Zhao Z, Huang X, Bai X, et al. Human exposure to phthalate esters associated with e-waste dismantling: Exposure levels, sources, and risk assessment. Environment International. 2019;**124**:1-9

[13] Lin M, Tang J, Ma S, Yu Y, Li G, Fan R, et al. Insights into bio monitoring of human exposure to polycyclic aromatic hydrocarbons with hair analysis: A case study in e-waste recycling area. Environment International. 2019;**136**:105432

[14] Awasthi AK, Wang M, Awasthi MK, Wang Z, Li J. Environmental pollution

and human body burden from improper recycling of e-waste in China: A short-review. Environmental Pollution. 2018; 243, Part B, 1310-1316

[15] Wang Y, Sun X, Fang L, Li K, Yang P, Du L, et al. Genomic instability in adult men involved in processing electronic waste in Northern China. Environment International. 2018;**117**:69-81

[16] Li J, Li W, Gao X, Liu L, Shen M, Chen H, et al. Occurrence of multiple classes of emerging photo initiators in indoor dust from E-waste recycling facilities and adjacent communities in South China and implications for human exposure. Environment International. 2020;**136**:105462

[17] Julander A, Lundgren L, Skare L, Grancér M, Palm B, Vahter M, et al. Formal recycling of e-waste leads to increased exposure to toxic metals: An occupational exposure study from Sweden. Environmental International. 2014;**73**:243-251

[18] Grant K, Coldizen FC, Sly PD, Brune MB, Neira M, Van den Berg M, et al. Health consequences of exposure to e-waste: A systematic review. Lancet Globalization and Health. 2013;**1**:350-361

[19] Heeks R, Subramanian L, Jones C. Understanding e-waste management in developing countries: Strategies, determinants, and policy implications in the Indian ICT Sector. Information Technology for Development. 2015;**21**(4):653-667. DOI: 10.1080/02681102.2014.886547

[20] Wang F, Kuehr R, Ahlquist D, Li J. E-waste in China: A country report. StEP Green Paper Series. 2013

[21] Zhang K, Schnoor JL, Zeng EY. E waste recycling: Where does it go from here? Environmental Science & Technology. 2022;**46**:10861-10867

[22] Karishma Chaudhary K. Case study analysis of e-waste management systems in Germany, Switzerland, Japan and India: A RADAR chart approach. Benchmarking An International Journal. 2018;**25**(9):3519-3540. DOI: 10.1108/BIJ-07-2017-0168

[23] UNFCCC. Greenhouse gas emissions reduction by recovering metals and materials through electronic waste collection and recycling process performed at Attero recycling Pvt Ltd plant located in Roorkee, Uttarakhand, India. Validation report. 2013. Available from: http://cdm.unfccc.int/filestorage/p/w/5NZ1O0LWF9DCX6S3HPGRQAYVJBT7MI.pdf/FVR_12

[24] Khetriwal DS, Kraeuchi P, Widmer R. Producer responsibility for e-waste management: Key issues for consideration—Learningfrom the Swiss experience. Journal of Environmental Management. 2009;**90**(1):153-165

[25] Wager PA, Hischier R, Eugster M. Environmental impacts of the Swiss collection and recovery systems for waste electrical and Electronic equipment (WEEE): A follow-up. Science of The Total Environment. 2011;**409**(10):1746-1756. DOI: 10.1016/j.scitotenv.2011.01.050

[26] Islam MT, Dias P, Huda N. Comparison of e-waste management in Switzerland and in Australia: A qualitative content analysis. International Journal of Environmental and Ecological Engineering. 2018;**12**(10):610-616

[27] EPI: Environmental Performance Index. 2018

[28] Bo B, Yamamoto K. Characteristics of e-waste recycling systems in

Japan and China. World Academy of Science. Engineering and Technology. 2010;**38**(2):500-506

[29] Abalansa S, El Mahrad B, Icely J, Newton A. Electronic waste, an environmental problem exported to developing countries: The GOOD, the BAD and the UGLY. Sustainability. 2021;**13**(9):5302. DOI: 10.3390/su13095302

[30] Rode S. E-Waste Management in Mumbai Metropolitan Region: Constraints and Opportunities. Theoretical and Empirical Researches in Urban Management. Research Centre in Public Administration and Public Services, Bucharest, Romania. 2012;7(2):90-104

[31] Sothun C. Situation of e-waste management in Cambodia. Procedia Environmental Sciences. 2012;**16**:535-544

[32] Chibunna JB, Siwar C, Begum RA, Mohamed AF. The challenges of e-waste management among institutions: A case study of UKM. Procedia - Social and Behavioral Sciences. 2012;**59**:644-649

[33] Panambunan-Ferse M, Breiter A. Assessing the side-effects of ICT development: Ewaste production and management a case study about cell phone end-of-life in Manado, Indonesia. Technology in Society. 2013;**35**(3):223-231

[34] Raha UL. E-waste management in Bangladesh: An overview. In: Conference Paper. International Conference on Urban and Regional Planning. Bangladesh: Bangladesh Institute of Planners; 2021

[35] Department of Environmental Affairs and Tourism. National Environmental Management: Waste Bill. Republic of South Africa: Department of Environmental Affairs and Tourism; 2007

[36] Awasthi AK, Zeng X, Li J. Comparative examining and analysis of E-waste recycling in typical developing and developed countries. Procedia Environmental Sciences. 2016;**35**:676-680

[37] Sudipta C, Javed SA, Biswa P. Reusing and recycling practice of e-waste in Dhaka city. In: International Conference on Engineering Research, Innovation and Education (ICERIE), Bangladesh; 2017. pp. 1-7

[38] ESDO. Magnitude of the Flow of E-Waste in Bangladesh. Dhaka, Bangladesh: ESDO; 2014

[39] Economic Times. Delhi-NCR May Generate 1 Lakh Metric Tonnes of E-Waste Per Annum: ASSOCHAM. India: Economic Times; 2016. Available from: www.assocham.org/upload/news/1461337882.pdf

[40] MoEF. E-waste (Management) Rules. India: MoEF; 2016. Available from: www.moef.gov.in

[41] Lu C, Zhang L, Zhong Y, Ren W, Tobias M, Mu Z, et al. An overview of e-waste management in China. Journal of Material Cycles and Waste Management. 2015;**17**:1-12. DOI: 10.1007/s10163-014-0256-8

[42] Wei L, Liu Y. Present status of e-waste disposal and recycling in China. Procedia Environmental Sciences. 2012;**16**:506-514

[43] Yoshida A, Terazono A, Ballesteros FC Jr, Nguyen D, Sukandar S, Kojima M, et al. E-waste recycling processes in Indonesia, the Philippines, and Vietnam: A case study of cathode ray tube TVs and monitors. Resources, Conservation and Recycling. 2016;**106**:48-58

[44] Ilankoon IMSK, Ghorbani Y, Chong MN, Herath G, Moyo T, Petersen J. E-waste in the international context – A review of trade flows, regulations, hazards, waste management strategies and technologies for value recovery. Waste Management. 2018;**82**:258-275

[45] Pandebesie ES, Indrihastuti I, Wilujeng AA, Warmadewanthi I. Factors influencing community participation in the management of household electronic waste in West Surabaya, Indonesia. Environmental Science and Pollution Research. 2019;**26**:27930-27939